WITHDRAWN

SUCCEED
IN MATHS

Editor:
Brian Seager

D0610679

HODDER
EDUCATION
PART OF HACHETTE LIVRE UK

Hachette's policy is to use papers that are natural, renewable and recyclable products and made from wood grown in sustainable forests. The logging and manufacturing processes are expected to conform to the environmental regulations of the country of origin.

Orders: please contact Bookpoint Ltd, 130 Milton Park, Abingdon, Oxon OX14 4SB. Telephone: (44) 01235 827720. Fax: (44) 01235 400454. Lines are open 9.00–5.00, Monday to Saturday, with a 24-hour message answering service. Visit our website at www.hoddereducation.co.uk

© Howard Baxter, Mike Handbury, John Jeskins, Jean Matthews, Mark Patmore, Brian Seager and Eddie Wilde © 2003, 2008
First published in 2003 © by Hodder Education,
part of Hachette Livre UK,
338 Euston Road,
London NW1 3BH.

This second edition first published 2008

Impression number   10 9 8 7 6 5 4 3 2 1
Year   2013 2012 2011 2010 2009 2008

Cover photo: ©Bernd Heusing/www.iStockphoto.com
Typeset in Rotis Serif 9.5pt on 11pt by Pantek Arts Ltd, Maidstone, Kent.
Printed in Spain

A catalogue record for this title is available from the British Library

ISBN-13: 978 0340 967775

CONTENTS

	Introduction	1
1	Percentages	2
2	Fractions	4
3	Non-calculator work	6
4	Ratio	8
5	Direct proportion, best buy and money problems	10
6	Primes, HCF, LCM and power laws	12
7	Estimation and rounding	14
8	Using a calculator	16
9	Algebraic manipulation	18
10	Substitution	20
11	Linear equations	22
12	Trial and improvement	24
13	Drawing graphs	26
14	Angle properties	28
15	Circle angle facts	30
16	Constructions and loci	32
17	Transformations	34
18	Pythagoras	36
19	Area, volume and perimeter	38
20	Speed and other compound units	40
21	Scatter diagrams	42
22	Cumulative frequency and box plots	44
23	Frequency diagrams and polygons	46
24	Stem-and-leaf diagrams	48
25	Pie charts	50
26	Basic probability	52
27	Tree diagrams	54
28	Further algebra	56
29	Repeated change and compound interest	58
30	Standard index form	60
	Answers	62

This book is intended to help you improve your grade if you are entered for the Higher tier. It can be used with any GCSE specification as they all derive from the same National Curriculum. It is not a complete revision guide and does not cover all the topics that will occur in a Higher paper. But it does target those topics where you are most likely to score marks in the examination.

Higher papers have questions that are targeted at grades D to A* but there are relatively more at the lower grades, so there will be plenty for you to attempt if you are aiming at grade C. However, it will not be sufficient just to get grade C questions right. You must get most of the easier grade D questions right too. Hopefully, you will also be successful with some of the grade B work covered in this book.

The topics selected from the specification cover all of the groups: Number; Algebra; Shape, space and measures; Handling data. Each topic is presented on two facing pages. The work is covered at two levels: *Basic* covers grade D and easier grade C, and *Moving on* harder grade C and grade B work. The *Moving on* work is on a blue background. It is a good idea to tackle the *Basic* work first. It is the questions on these you will need to get right. When you are confident with these, move on! In each section, there are reminders ❗ about what you should know and worked examples to show you how to answer questions. These are followed by practice questions for you to try to see if you have understood. You also have answers to check whether you are right!

Some topics are intended to be done without a calculator. The questions here are marked with ⌛ and you may expect them to appear in the non-calculator section/paper in the examination.

When you have worked through this book you should be able to do all the easy questions without making silly mistakes. You should also be able to do some of the harder questions. Don't waste time trying questions you do not feel you can do until you have attempted all the others carefully.

Examination tips

There will be a formula sheet in the examination. Amongst other things it will have the area of a trapezium and the volume of a prism. Any others covered in this book you will have to learn.

Read the instructions carefully, both on the front of the paper and for each question. Here are some of the words that will be used and what they mean.

Write down, state – little working out will be needed and no explanation.
Calculate, find – something to work out. Show all the steps of your working.
Solve – show all the steps in solving your equation.
Show – give all the steps and reasons.
Draw – draw as accurately as you can.
Sketch – need not be accurate but show all the features.
Explain – give brief reasons.   Brian Seager (*Editor*) 2008

PERCENTAGES

$35\% = \frac{35}{100} = 0.35$ $20\% = \frac{20}{100} = 0.2$ $6\% = \frac{6}{100} = 0.06$ etc.

Example Work out 30% of £60.

Solution 30% of £60 = 0.3×60 = £18

> **TIP!** Take care with the decimal point.

To find A as a percentage of B, divide A by B and multiply by 100.

Example Barry scored 24 out of 60 in a test. What percentage was this?

Solution 24 out of 60 = $\frac{24}{60} \times 100\% = 0.4 \times 100\% = 40\%$

To increase or decrease by 5%, say, add to or subtract from 100% to get 105% or 95%. Multiply by 1·05 or 0·95.

Example A computer is priced at £720 plus VAT. Vat is 17·5%. What is the total price of the computer?

Solution Cost before VAT = 100%
Cost including VAT = 100 + 17·5% = 117·5%
Total price = 1.175×720 = £846

Percentage increase or decrease = $\frac{\text{(increase or decrease)}}{\text{original amount}} \times 100$

Example Lisa bought a car for £7500 and sold it two years later for £4900. What % loss did she make? Give the answer to 1 decimal place.

Solution Loss = £2600
Percentage loss = $\frac{2600}{7500} \times 100$
= 34·666 = 34·7%

> **TIP!** Subtract to find loss.
> **TIP!** Divide the loss by the original.

If 5% is added each year, multiply each year's total by 1·05.

Example Sheila invested £1500 for 3 years with 6% interest added each year. How much did she have at the end of that time? Give the answer to the nearest penny.

Solution Increase = 6% per year
Amount after 1 year = 106% of original
= 1.06×1500 = £1590
Amount after 2 years = 1.06×1590 = £1685·40
Amount after 3 years = 1.06×1685.40
= £1786·524 = £1786·52

> **TIP!** Work out each year separately.

Practice questions

1 Work out these.
   (a) 30% of £20 **(b)** 40% of £15 **(c)** 25% of £90 **(d)** 60% of £240

2 What percentage are each of these?
   (a) 20 of 80   **(b)** 30 of 200   **(c)** 45 of 150
   (d) 72 of 120 **(e)** 20 of 400

3 In a school of 500 pupils 45% are boys. How many boys are there?

4 In a raffle 350 out of the 400 tickets sold. What percentage were sold?

5 In a sale all items were reduced by 8%. A coat was priced at £80 before the sale. What was the price in the sale?

6 To laminate a large poster the cost was £24 plus VAT at 17·5%. What was the total cost?

7 Joe spent £1600 on his holiday. Of this 60% was for accommodation. How much did he pay for accommodation?

8 In a marathon race 850 out of the 1250 competitors finished in under three hours. What percentage was this?

9 Iain bought a caravan for £7500 and sold it at a profit of 4%. What price did he sell it for?

10 On a Saturday at Cluster Park 170 of the 240 people who entered the park were children.
    What percentage were children? Give the answer to 1 decimal place.

11 Pam bought a coat for £60 and sold it later for £45. What percentage loss did she make?

12 The cost price of a scooter was £120 and the selling price was £144. What was the percentage profit?

13 Jack bought a chair for £90 and sold it at a loss of 10%. What price did he sell it for?

14 The following are cost prices and selling prices of a number of items. For each find the percentage profit or loss.
    (a) Cost price = £15   Selling price = £15·45
    (b) Cost price = £16   Selling price = £14

15 David invested £25 000 for three years and at the end of each year 4·5% was added to the total. How much did he have at the end of the three years? Give the answer to the nearest pound.

16 The value of Paul's car dropped by 10% each year. He bought it for £17 500. What was its value at the end of four years?

17 Antionette's salary was £15 000 in 1994. She was given a 3% increase each year for the next three years. What was her salary at the end of the three years? Give the answer to the nearest penny.

FRACTIONS

Comparing fractions

There are two main ways to compare fractions.
- Change them into decimals.
- Change them into fractions which have the same denominator.

Example Which of $\frac{3}{8}$ and $\frac{2}{5}$ is bigger?

Solution
$$8\overline{)3 \cdot 0\,0\,0}^{\,0 \cdot\,3\,7\,5} \qquad 5\overline{)2 \cdot 0}^{\,0 \cdot\,4} \qquad \frac{2}{5} \text{ is bigger.}$$

or
$$\frac{3}{8} \times \frac{5}{5} = \frac{15}{40} \qquad \frac{2}{5} \times \frac{8}{8} = \frac{16}{40} \qquad \frac{2}{5} \text{ is bigger.}$$

> **TIP!**
> Remember that $\frac{3}{8} = 3 \div 8$.

Adding and subtracting fractions

Fractions must have the same denominator before they can be added or subtracted. Then just add or subtract the numerators **only**.

Example Work out these. (a) $\frac{2}{3} + \frac{3}{10}$ (b) $\frac{11}{12} - \frac{3}{4}$

Solution (a) $\frac{20}{30} + \frac{9}{30} = \frac{29}{30}$ (b) $\frac{11}{12} - \frac{9}{12} = \frac{2}{12}$

Adding and subtracting mixed numbers

- Add the whole numbers and add the fractions separately.
- When subtracting, if the first fraction is smaller, add a whole number to it.

Example Work out these. (a) $2\frac{2}{3} + 1\frac{3}{8}$ (b) $4\frac{1}{4} - 1\frac{2}{5}$

Solution (a) $2 + 1 + \frac{2}{3} + \frac{3}{8} = 3 + \frac{16}{24} + \frac{9}{24} = 3 + 1\frac{1}{24} = 4\frac{1}{24}$

(b) $3 - 1 + 1\frac{1}{4} - \frac{2}{5} = 2 + \frac{5}{4} - \frac{2}{5} = 2 + \frac{25}{20} - \frac{8}{20} = 2\frac{17}{20}$

Multiplying fractions

Cancel first if you can. Then multiply tops and bottoms separately.

Example Work out these. (a) $\frac{3}{8} \times \frac{4}{5}$ (b) $\frac{3}{7} \times 4$

Solution (a) $\frac{3}{{}_2 8} \times \frac{4^1}{5} = \frac{3}{10}$ (b) $\frac{3}{7} \times \frac{4}{1} = \frac{12}{7} = 1\frac{5}{7}$

> **TIP!**
> Divide out top heavy fractions.

Dividing fractions

Turn the **second** fraction upside down, cancel and then multiply tops and bottoms.

Example Work out these. (a) $\frac{3}{5} \div \frac{9}{10}$ (b) $\frac{3}{4} \div 2$

Solution (a) $\frac{{}^1 3}{{}_1 5} \times \frac{10^2}{9_3} = \frac{2}{3}$ (b) $\frac{3}{4} \div \frac{2}{1} = \frac{3}{4} \times \frac{1}{2} = \frac{3}{8}$

> **TIP!**
> The fraction in part (a) is in its **simplest form** or **lowest terms**.

Practice questions

1 In each of these parts, work out which of the two fractions is bigger.

    (a) $\frac{3}{4}, \frac{5}{8}$ **(b)** $\frac{2}{9}, \frac{1}{4}$ **(c)** $\frac{3}{5}, \frac{7}{12}$ **(d)** $\frac{4}{15}, \frac{3}{10}$

2 Add these fractions.

    (a) $\frac{1}{4} + \frac{5}{8}$ **(b)** $\frac{1}{2} + \frac{1}{6}$ **(c)** $\frac{2}{3} + \frac{1}{4}$ **(d)** $\frac{1}{2} + \frac{2}{5}$

    (e) $\frac{3}{5} + \frac{1}{10}$ **(f)** $\frac{7}{10} + \frac{1}{4}$ **(g)** $\frac{1}{6} + \frac{11}{15}$ **(h)** $\frac{4}{5} + \frac{1}{3}$

3 Subtract these fractions.

    (a) $\frac{2}{3} - \frac{1}{6}$ **(b)** $\frac{3}{4} - \frac{5}{12}$ **(c)** $\frac{5}{6} - \frac{1}{3}$ **(d)** $\frac{11}{12} - \frac{2}{3}$

    (e) $\frac{3}{4} - \frac{5}{7}$ **(f)** $\frac{9}{10} - \frac{3}{4}$ **(g)** $\frac{11}{15} - \frac{2}{5}$ **(h)** $\frac{8}{9} - \frac{5}{6}$

4 Work out these.

    (a) $1\frac{1}{2} + 2\frac{1}{3}$ **(b)** $4\frac{1}{2} - 2\frac{1}{3}$ **(c)** $6\frac{2}{5} + 3\frac{1}{2}$ **(d)** $1\frac{3}{4} + 2\frac{2}{3}$

    (e) $3\frac{8}{15} - 2\frac{1}{3}$ **(f)** $7\frac{2}{5} - 4\frac{5}{6}$ **(g)** $1\frac{1}{3} + 4\frac{6}{7}$ **(h)** $3\frac{3}{4} - 2\frac{4}{5}$

5 Multiply these fractions.

    (a) $\frac{3}{5} \times 6$ **(b)** $11 \times \frac{2}{3}$ **(c)** $\frac{1}{5} \times \frac{2}{5}$ **(d)** $\frac{3}{8} \times \frac{7}{10}$

    (e) $\frac{1}{6} \times \frac{3}{4}$ **(f)** $\frac{3}{5} \times \frac{4}{9}$ **(g)** $\frac{5}{12} \times \frac{3}{10}$ **(h)** $\frac{4}{15} \times \frac{5}{8}$

6 Divide these fractions.

    (a) $\frac{1}{2} \div 3$ **(b)** $4 \div \frac{1}{3}$ **(c)** $\frac{1}{4} \div \frac{2}{5}$ **(d)** $\frac{2}{7} \div \frac{5}{9}$

    (e) $\frac{1}{4} \div \frac{5}{6}$ **(f)** $\frac{1}{2} \div \frac{3}{10}$ **(g)** $\frac{3}{5} \div \frac{9}{10}$ **(h)** $\frac{4}{15} \div \frac{2}{9}$

7 Work out these.   **(a)** $\frac{1}{4} + \frac{1}{6} + \frac{5}{12}$ **(b)** $\frac{2}{3} + \frac{1}{5} - \frac{7}{15}$

                                   (c) $\frac{3}{15} \times \frac{4}{9} \times \frac{5}{8}$ **(d)** $\frac{5}{12} \times \frac{3}{4} \div \frac{5}{16}$

NON-CALCULATOR WORK

Multiplying and dividing by 10, 100, 1000, ...
- When multiplying by 10, 100, etc. the digits stay in the same order.
- To multiply, move the decimal point to the right.
- To divide, move the decimal point to the left.

Example Multiply 23·4 by 100.

Solution $23·4 \times 100 = 2340$

TIP!
$23 \cdot \widehat{40}$

Example Divide 57 by 1000.

Solution $57 \div 1000 = 0·057$

TIP!
The decimal point is after the 7 in 57.

Adding and subtracting decimals
- Always set out addition and subtraction down the page, **not** across the page.
- Keep the decimal points underneath each other.

Example Work out 14·6 + 5·27.

Solution
```
    14·6
 +  5·27
   19·87
```

TIP!
Put a point after the 12 and fill empty decimal places with 0s.

Example Work out 12 – 3·86.

Solution
```
 12¹·⁰0¹0
 − 3·8 6
   8·1 4
```

Multiplication
- $0·2 \times 0·3 = 0·06$ as $2 \times 3 = 6$ and there are two decimal places.
- $200 \times 30 = 6000$ as $2 \times 3 = 6$ and there are three zeros.

Example Work out these. **(a)** $(0·3)^2$ **(b)** $0·08 \times 600$

Solution **(a)** $3 \times 3 = 9$ so $0·3 \times 0·3 = 0·09$
(b) $8 \times 6 = 48$ so $0·08 \times 600 = 48$

TIP!
Two decimal places cancel the two 0s.

Dividing by a whole number
When dividing by a whole number, keep the decimal point lined up.

Example Work out 146·96 ÷ 8.

Solution
```
      1 8. 3 7
  8 ⟌ 14⁶6·⁷9 ⁶6
```

Percentages
A quick way to find 15% is to find 10%, then 5% and add.

Example Find 35% of £8.

Solution 10% of £8 = £0·80, 30% of £8 = $3 \times 0·80$ = £2·40
5% of £8 = $\frac{1}{2} \times 0·80$ = £0·40, so 35% of £8 = £2·40 + £0·40 = £2·80

Practice questions

1 Work out these.
   (a) $14·52 \times 10$ **(b)** $17·3 \times 100$ **(c)** $24·6 \div 10$
   (d) $27 \div 100$ **(e)** $0·46 \times 1000$ **(f)** $1·72 \div 100$
   (g) $29·32 \times 10\,000$ **(h)** $836 \div 1000$ **(i)** $28·69 \times 100\,000$

In the following seven questions you will need these conversions. (You should learn them for GCSE.)
     $1\,kg = 1000\,g$ $1\,litre = 1000\,ml$ $1\,cm = 10\,mm$ $1\,m = 100\,cm$

2 Convert $7·3\,kg$ to g.

3 Convert $6340\,g$ to kg.

4 Convert $2·78\,litres$ to ml.

5 Convert $738\,ml$ to litres.

6 Convert $0·37\,cm$ to mm.

7 Convert $7·8\,m$ to cm.

8 Convert $435\,cm$ to m.

9 Work out these.
   (a) $32·7 + 3·6 + 8·32$ **(b)** $56·38 + 7·96 + 6·483$
   (c) $27·89 - 8·3$ **(d)** $54·3 - 26·58$

10 Work out these.
   (a) $317·1 \div 7$ **(b)** $4·504 \div 8$ **(c)** $0·1176 \div 6$ **(d)** $62·76 \div 12$

11 Find the perimeter of this triangle.

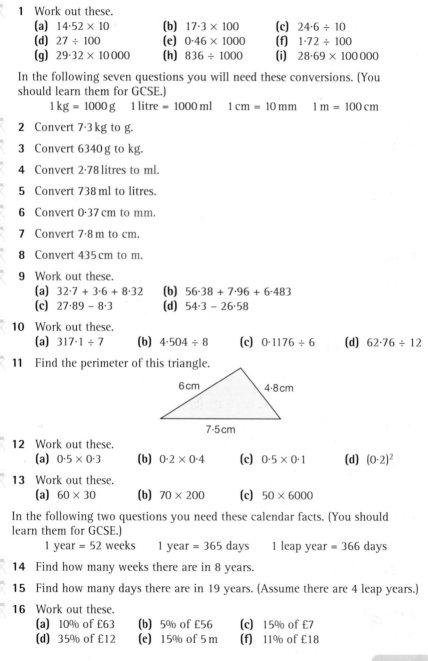

12 Work out these.
   (a) $0·5 \times 0·3$ **(b)** $0·2 \times 0·4$ **(c)** $0·5 \times 0·1$ **(d)** $(0·2)^2$

13 Work out these.
   (a) 60×30 **(b)** 70×200 **(c)** 50×6000

In the following two questions you need these calendar facts. (You should learn them for GCSE.)
     1 year = 52 weeks   1 year = 365 days   1 leap year = 366 days

14 Find how many weeks there are in 8 years.

15 Find how many days there are in 19 years. (Assume there are 4 leap years.)

16 Work out these.
   (a) 10% of £63   **(b)** 5% of £56   **(c)** 15% of £7
   (d) 35% of £12   **(e)** 15% of 5 m   **(f)** 11% of £18

RATIO

> To divide an amount in the ratio of 5 : 4, divide the original amount by 9, (5 + 4), and multiply by 5 for the first share and by 4 for the second.

Example   Divide £450 between Jodi and Helen in the ratio 2 : 3.

Solution   2 + 3 = 5, £450 ÷ 5 = £90
Therefore Jodi receives 2 × £90 = £180 and
Helen receives 3 × £90 = £270.
Check £180 + £270 = £450

> **TIP!** It is a good idea to check that the shares add up to the total.

> With more than two shares the method is the same.

Example   Jodi, Helen and Kathy share £1500 in the ratio of 10 : 8 : 7. How much does each get?

Solution   10 + 8 + 7 = 25, £1500 ÷ 25 = £60
Jodi gets 10 × £60 = £600, Helen gets 8 × £60 = £480,
Kathy gets 7 × £60 = £420.
Check £600 + £480 + £420 = £1500

> Ratios can be used to calculate increases and decreases. Change the ratio into a multiplying factor.

Example   The length of a photograph is 36 cm. The photograph is enlarged in the ratio of 3 : 4. Calculate the enlarged length.

Solution   3 : 4 gives a multiplying factor of $\frac{4}{3}$
The enlarged length = $36 \times \frac{4}{3} = 48$ cm

> Ratios are used with scale drawings and maps.

> **TIP!** 1 m = 100 cm
> 1 km = 1000 m

Example   Two towns are 10 cm apart on a map of scale 1 : 25 000. What is the actual distance between the towns?

Solution   1 cm on the map = 25 000 cm on the ground.
Therefore 10 cm on the map = 250 000 cm on the ground
= 2·5 km

Example   The scale of a model pier is 1 : 100.
(a) The full size pier is 230 m long. How long is the model pier?
(b) On the model, the ballroom is 35 cm long. What is the length of the real ballroom in metres?

Solution   (a) 230 ÷ 100 = 2·3 m
(b) 35 × 100 ÷ 100 = 35 m

Practice questions

1 Divide £56 between three people in the ratio $2:3:3$.

2 Decrease the length of a line 25 cm long in the ratio $5:3$.

3 An architect makes a model of a building using a scale of $1:20$. The front of the building is 30 metres long. How long is the front of the model building?

4 The length of a road is 2·8 km. How long will the road be on a map of scale $1:10\,000$?

5 The scale of a drawing is $1:1000$. The length of a wall is 25 m. What length will the wall be on the drawing?

6 In an election 28 000 people voted Labour, Conservative or Liberal Democrat in the ratio $7:5:2$. How many people voted for the Liberal Democrats?

7 In a class the ratio of boys to girls is $3:4$. If there are 12 boys, how many girls are there?

8 A photographer enlarges an original picture in the ratio $2:3$. The height of a tree in the original picture is 13 cm. How tall is the tree on the enlarged copy?

9 Share £30 in the ratio $2:3:5$.

10 A picture is 40 cm long and 30 cm wide. A photographic enlargement is to be made for an advertisement. The length of the enlargement is to be 100 cm. What is the width of the enlargement?

11 Here is a recipe for flapjacks:

> l00 g of butter
> 75 g of syrup
> 75 g of sugar
> 200 g of oats

The ratio of butter to oats is $100\,g:200\,g$ or $1:2$ in its simplest form.
(a) What is the ratio of sugar to oats? Give your answer in its simplest form.
This recipe will make 25 flapjacks.
(b) How much butter is needed for 40 flapjacks?

12 A photocopier reduces in the ratio $5:3$.
The height of a church spire on an original picture is 12 cm.
How high is the spire on the reduced picture?

13 Tom, Dick and Harry share £1080 in the ratio $2:4:6$.
How much does Dick receive?

14 In a shop the ratio of oranges to apples is $2:5$. If there are 60 apples, how many oranges are there?

15 The distance between two towns is 18 km.
How far apart will they be on a map with a scale of $1:50\,000$?

DIRECT PROPORTION, BEST BUY AND MONEY PROBLEMS

> **Direct proportion**
> If y is proportional to x ($y \propto x$):
> - If you double x you double y.
> - If you treble x you treble y and so on.
> - The equation connecting x and y is $y = kx$.

Example   The mass of a rope is proportional to the length. A 50 metre rope weighs 60 kg.
(a) How much will a 75 metre rope weigh?
(b) What length of rope will weigh 150 kg?

Solution   (a) 75 is $1\frac{1}{2} \times 50$ so weight $= 1\frac{1}{2} \times 60 = 90$ kg
(b) $150 \div 60 = 2\cdot5$ so length $= 50 \times 2\cdot5 = 125$ m

> If y is proportional to x^2 ($y \propto x^2$):
> - Write the equation $y = kx^2$.
> - Substitute in a pair of corresponding values of x and y to find k.

Example y is proportional to x^2 and $y = 20$ when $x = 2$.

(a) Find an equation connecting x and y.

(b) Find the value of x when $y = 1\cdot4$.

Solution   (a) $y = kx^2$. Substitute $x = 2$ and $y = 20$ to give $20 = k \times 4$.
So $k = 5$ and $y = 5x^2$
(b) $y = 5 \times 1\cdot4^2 = 9\cdot8$

> **Finding which is better value**
> - Compare the same quantities.
> - Write clearly what information your calculations give.

Example   Which of these offers is better value?

2 litre bottles	1·5 litre bottles
Special offer	5̶0̶p̶
3 for £2	**Reduced price** 45p

Solution
6 litres cost £2   1·5 litres cost 45p
1 litre costs £2 $\div$ 6 $= 33\cdot3...$p   1 litre costs $45 \div 1\cdot5 = 30$p

The price per litre is cheaper with the 1·5 litre bottle.

Practice questions

1 y is proportional to x.
Copy and complete this table.

x	3	5	8
y	15	25	40

2 $W \propto A$.
Copy and complete this table.

A	10	8	4
W	25	20	10

3 A car goes 258 miles on 6 gallons of petrol. The distance (d miles) is proportional to the number of gallons (g) used.
(a) Find an equation connecting d and g.
(b) Find the distance the car will go on 8 gallons of petrol.
(c) Find the number of gallons needed to travel 500 miles. Give the answer to 1 decimal place.

4 At one time the cost of a journey by train was proportional to the distance travelled. A journey of 200 miles cost £8.
(a) Find the cost for a journey of 80 miles.
(b) Find how far you could go for £5.

5 y is proportional to x^2. When $x = 3$, $y = 36$.
(a) Find an equation connecting x and y.
(b) Find y when $x = 5$.
(c) Find x when $y = 400$.

6 When a stone is dropped from a cliff, the distance (d metres) it falls is proportional to the square of the time (t seconds) it has been falling. After 2 seconds it has fallen 20 m.
(a) Find an equation connecting d and t.
(b) Find the distance it falls in 4 seconds.
(c) Find the time it takes to fall 125 m.

7 Which is better value?

8 A 750 g piece of cheese costs £3·96. Find the cost of a 1·250 kg piece of this cheese.

9 Ahmed buys a computer priced £756 on interest free credit. He pays 15% deposit, then 18 monthly payments. Calculate his monthly payments.

10 One year Petra earned £11 750. She had a tax-free allowance of £4500 and paid tax at 22% on the rest of her earnings. How much tax did she pay that year?

11 There are two interest free schemes for buying a car at one garage.
Scheme A: deposit 20%, then pay over 2 years.
Scheme B: deposit 30% then pay over 3 years.
Calculate the deposit and monthly payment for each scheme for a car priced £8460.

PRIMES, HCF, LCM AND POWER LAWS

> • A **prime number** has two different factors, itself and 1.
> • A **prime factor** is a factor that is prime.

Example Write 84 in prime factor form.

Solution

$$84 = 2^2 \times 3 \times 7$$

> TIP!
> Draw a factor tree to help you.

> The **Highest Common Factor (HCF)** of two or more numbers is the largest number which is a factor of all of them.

Example Find the HCF of 16 and 56.

Solution Factors of 16 are 1, 2, 4, 8, 16.
Factors of 56 are 1, 2, 4, 7, 8, 14, 28, 56.
The HCF of 16 and 56 is 8.

> The **Lowest Common Multiple (LCM)** of two or more numbers is the smallest number into which they will all divide.

Example Find the LCM of 24 and 42.

Solution $24 = 2 \times 2 \times 2 \times 3$ $42 = 2 \times 3 \times 7$
The LCM of 24 and 42 is $2 \times 2 \times 2 \times 3 \times 7 = 168$.

> TIP!
> Take the largest power of each factor.

> ### Rules for indices
> When combining either letters or numbers which have powers, you can use the following rules:
> • $x^a \times x^b = x^{a+b}$
> • $x^a \div x^b = x^{ab}$
> • $(x^a)^b = x^{a \times b}$
> Also $x^1 = x$ and $x^0 = 1$.

Example Simplify each of these to a single power.

(a) $x^3 \times x^4$ (b) $5^7 \div 5^2$ (c) $(p^4)^5$

(d) $y^4 \times y \times y^8$ (e) $\dfrac{7^6 \times 7^4}{7^7 \times 7^3}$ (f) $5t^6 \times 3t^3 \times 2t^2$

Solution

(a) $x^{3+4} = x^7$ (b) $5^{7-2} = 5^5$ (c) $p^{4 \times 5} = p^{20}$

(d) $y^{4+1+8} = y^{13}$ (e) $\dfrac{7^{6+4}}{7^{7+3}} = \dfrac{7^{10}}{7^{10}} = 7^{10-10} = 7^0 = 1$

(f) $5 \times 3 \times 2 \times t^{6+3+2} = 30t^{11}$

> TIP!
> Combine the numbers and the letters separately.

Practice questions

1　Express these numbers in prime factor form.
　　(a) 16　　　　**(b)** 36　　　　**(c)** 45　　　　**(d)** 210　　　**(e)** 300

2　Find the Highest Common Factor of these numbers.
　　(a) 28 and 105　　**(b)** 56 and 248　　**(c)** 78 and 325
　　(d) 63 and 88　　　**(e)** 24, 42 and 56　　**(f)** 36, 126 and 243
　　(g) 34, 102 and 153

3　Find the Lowest Common Multiple of these numbers.
　　(a) 36 and 63　　**(b)** 28 and 35　　　**(c)** 24 and 54
　　(d) 42 and 64　　**(e)** 14, 35 and 40　　**(f)** 18, 42 and 66
　　(g) 24, 42 and 56

4　Minties are available in packets and in boxes. Packets of Minties contain 18 sweets and boxes of Minties contain 32 sweets. What is the smallest number of sweets needed so that a packer could fill either packets only or boxes only without having any sweets left over?

5　Barnies pack eggs in cartons of 6, 7, 9, 10 and 12. What is the smallest number of eggs that would be needed so that all the different carton sizes could be filled, regardless of which size of carton is chosen, without leaving any spare eggs?

In the following questions, simplify each expression to a single power.

6　**(a)** $3^2 \times 3^4$　　　　　**(b)** $a^4 \times a^5$

7　**(a)** $5^6 \div 5^4$　　　　　**(b)** $\dfrac{9^7}{9^5}$　　　　　**(c)** $k^8 \div k^7$

8　**(a)** $(p^2)^3$　　　　　**(b)** $(7^5)^6$

9　**(a)** $x \times x^8$　　　　**(b)** $y^2 \times y^3 \times y^4$　**(c)** $2^4 \times 2^2 \times 2^3 \times 2^6$

10　**(a)** $p^5 \times p^2 \times p^8 \times p^7$　　　　　**(b)** $4t^4 \times 6t^3 \times 5t^4$

11　**(a)** $\dfrac{15x^6}{3x^3}$　　　　**(b)** $\dfrac{y^9 \times y^4}{y^5 \times y^3}$

12　**(a)** $8m^4 \times m^2 \times 4m$　**(b)** $(q^6)^0$　　　**(c)** $4t \times 3t^3 \times 7t^0$

13　**(a)** $\dfrac{7^5 \times 7^4}{7^3}$　　　　**(b)** $\dfrac{6x^2 \times 12x^4}{3x \times 3x^2}$　　**(c)** $(2p^4)^3$

ESTIMATION AND ROUNDING

> ## Rounding
> If the value of the first digit not required is 5 or more, round up.

Example Round these numbers to the degree of accuracy stated.
    (a) 264 (nearest 10)   **(b)** 6317 (2 significant figures)
    (c) 47·89 (1 decimal place)   **(d)** 0·016 54 (2 significant figures)

Solution   **(a)** 260   **(b)** 6300
    (c) 47·9   **(d)** 0·017

> **TIP!** Include zeros to show the size.

> **TIP!** The first significant figure is the first non-zero digit.

> ## Estimation
> Always round to 1 significant figure when estimating.

Example Estimate the value of $\dfrac{29 \times 170}{87}$

> **TIP!** Divide to 1 decimal place.

Solution Estimate of $\dfrac{29 \times 170}{87} = \dfrac{30 \times 200}{90}$

$$= \frac{200}{3} = 66 \cdot 6$$

$$= 67 \text{ or } 70$$

> **TIP!** Do not be too accurate.

Example Estimate the value of $\dfrac{4 \cdot 7 \times 30 \cdot 1}{(5 \cdot 7 - 1 \cdot 9)}$

Solution Estimate of $\dfrac{4 \cdot 7 \times 30 \cdot 1}{(5 \cdot 7 - 1 \cdot 9)} = \dfrac{5 \times 30}{(6 - 2)}$

$$= \frac{150}{4} = 37 \cdot 5$$

$$= 37, 38 \text{ or } 40$$

> **TIP!** The final answer should be to 1 or 2 s.f.

> When dividing by 0.2, multiply by 10 and divide by 2.

Example Estimate the value of
    (a) $\dfrac{612 \times 0 \cdot 27}{11 \cdot 6}$ **(b)** $\dfrac{39 \cdot 5 \times 41 \cdot 3}{0 \cdot 079}$

Solution   **(a)** $\dfrac{600 \times 0 \cdot 3}{10} = 18$

> **TIP!** Be careful about the decimals.

    (b) $\dfrac{40 \times 40}{0 \cdot 08} = \dfrac{1600}{0 \cdot 08} = \dfrac{160\,000}{8} = 20\,000$

> **TIP!** Multiply top and bottom by 100 to make 0·08 a whole number.

Practice questions

1 Round these numbers to the degree of accuracy stated.

    (a) 1254 (nearest 10)   **(b)** 1257 (nearest 100)   **(c)** 7214 (nearest 1000)

    (d) 3617 (2 s.f.)   **(e)** 7·183 (1 d.p.)   **(f)** 0·027 54 (1 s.f.)

    (g) 1·666 (2 d.p.)   **(h)** 0·0632 (2 s.f.)

2 The number of people who ran in the Robin Hood half marathon was 10 575. What was this to the nearest thousand?

3 Estimate the value of each of the following.

    (a) 47×81 **(b)** 205×59 **(c)** $\dfrac{19 \times 31}{48}$

    (d) $\dfrac{216 \times 19}{37}$ **(e)** $\dfrac{59 \times 503}{28 \times 52}$

4 Estimate the value of each of the following.

    (a) $\dfrac{2 \cdot 2 \times 17 \cdot 1}{4 \cdot 9}$ **(b)** $\dfrac{3 \cdot 07 + 5 \cdot 71}{9 \cdot 1 - 6 \cdot 8}$

    (c) $\dfrac{24 \cdot 1 \times 19 \cdot 8}{6 \cdot 04 - 2 \cdot 13}$ **(d)** $\dfrac{71 \cdot 3 \times 48 \cdot 9}{81 \cdot 2 - 49 \cdot 7}$

5 Work out the value of the expressions in question **4**, giving the answers to 1 decimal place.

6 Estimate the value of each of the following.

    (a) $412 \times 0 \cdot 37$ **(b)** $0 \cdot 842 \times 0 \cdot 0631$

    (c) $\dfrac{0 \cdot 47}{8 \cdot 14}$ **(d)** $\dfrac{2 \times 587 \cdot 6}{73 \cdot 2 + 18 \cdot 4}$

7 Estimate the value of each of the following.

    (a) $\dfrac{214 \times 187}{0 \cdot 76}$ **(b)** $\dfrac{512 + 692}{0 \cdot 79}$

    (c) $\dfrac{43 \cdot 6 \times 39 \cdot 1}{0 \cdot 812 \times 6 \cdot 31}$ **(d)** $\dfrac{2 \cdot 76 \times 594}{6 \cdot 83 \times 0 \cdot 711}$

8 Trevor drove 187 miles in 3 hours 47 minutes. Estimate his average speed.

9 **(a)** Without using a calculator, find an approximate value for the population density in people per km^2 of an island with an area of 52·9 km^2 and a population of 3217.

    (b) Now use your calculator to work out the value correct to 3 significant figures.

10 A cone has a base radius of 4·7 cm and a vertical height of 15·3 cm. The formula for the volume of a cone is $V = \frac{1}{3}\pi r^2 h$.

Work out the volume of the cone correct to 3 significant figures.

USING A CALCULATOR

Buttons on a calculator

Look for these buttons on your calculator:

- $($ and $)$ are brackets.
- $(-)$ or $+/-$ make negative numbers.
- x^2 squares a number.
- x^3 cubes a number.
- x^y or y^x or $^\wedge$ work out **any** power of a number (including squares and cubes).
- $\sqrt{}$ finds the square root of a number.
- $\sqrt[3]{}$ finds the cube root of a number.
- a^b/c works with fractions.
- $\boxed{\text{SHIFT}}$ or $2^{\text{nd}}\ \text{F}$ or $\boxed{\text{INV}}$ for the 'second function' of each button.

Using brackets

The calculator works out multiplications and divisions before additions and subtractions. Use brackets if this is not the required order.

Example   Work out $\dfrac{2814 - 513}{45}$

Solution

$(\quad 2 \quad 8 \quad 1 \quad 4 \quad - \quad 5 \quad 1 \quad 3 \quad) \quad \div \quad 4 \quad 5 \quad =$

Answer: 51·13 (to 2 decimal places).

> **TIP!**
> The top part should be worked out first, so put brackets round it.

Estimating answers

Estimate the answer **before** using the calculator, to give an idea of what answer to expect.

> **TIP!**
> Round each number in the calculation to 1 significant figure.

Example   Estimate the answer to $\dfrac{2814 - 513}{45}$

Solution $2814 \approx 3000$ $513 \approx 500$ $45 \approx 50$

Estimate: $\dfrac{2814 - 513}{45} = \dfrac{3000 - 500}{50} = \dfrac{2500}{50} = 50$

> **TIP!**
> The estimate and the actual answer should be quite close.

Example   Work out these.

(a) $-4{\cdot}2 - (-6{\cdot}9)$ (b) $\dfrac{1{\cdot}5 \times 4{\cdot}7 + 2{\cdot}8}{9{\cdot}3 \times 6{\cdot}6}$ (c) $\sqrt{9{\cdot}2^3 - 4{\cdot}5^2}$ (d) $\dfrac{2}{5} + 3\dfrac{4}{7}$

Solution   (a) $+/- \quad 4 \quad \cdot \quad 2 \quad - \quad +/- \quad 6 \quad \cdot \quad 9 \quad =$ Answer: 2·7

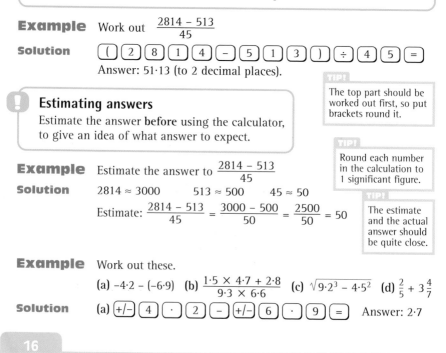

Practice questions

Work out these. Give the answers exact or to 2 decimal places.

1 $3 \cdot 8^2 - \sqrt{75}$

2 $\sqrt[3]{2197}$

3 $\dfrac{12 \cdot 4}{4 \cdot 23 - 1 \cdot 9}$

4 $\dfrac{15 \cdot 6 - 2 \cdot 7}{3 \cdot 2 \times 1 \cdot 4}$

5 $\dfrac{3 \cdot 7 + 4 \cdot 9}{8 \cdot 8 - 8 \cdot 5}$

6 $-6 \cdot 5(1 \cdot 2 + 4 \cdot 6)$

7 $(2 \cdot 3 - 1 \cdot 8)^2 \times 1 \cdot 07$

8 $11\frac{3}{4} - 2\frac{4}{7}$

9 $\sqrt{31 \cdot 9^2 + 8 \cdot 77^2}$

10 $\frac{4}{5} \div \frac{2}{3}$

11 $2 \cdot 5^4$

12 $\sqrt{\dfrac{8 \cdot 79}{0 \cdot 035}}$

13 $\sqrt[3]{3 \cdot 87}$

14 $5\frac{1}{4} \times 2\frac{3}{5}$

15 $200(1 + 0 \cdot 056)^8$

16 $\sqrt{\dfrac{5 \cdot 68^3}{4 \cdot 75 - (-2 \cdot 59)}}$

17 $\dfrac{2 \cdot 6 - 4 \cdot 3 \div 1 \cdot 5}{3 \cdot 3 \times 7 \cdot 1 + 9 \cdot 3}$

18 $\dfrac{4 \cdot 5 \sin 60^\circ}{\sin 30^\circ}$

19 $9^{1/3} \times 4^5$

20 $\dfrac{(22 \cdot 4 + 13 \cdot 2) \times 6 \cdot 3^2}{185 \cdot 9 \times (9 \cdot 5 - 3 \cdot 87)}$

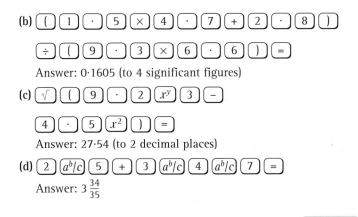

(b) $(\; 1 \; . \; 5 \; \times \; 4 \; . \; 7 \; + \; 2 \; . \; 8 \;)$

$\div \; (\; 9 \; . \; 3 \; \times \; 6 \; . \; 6 \;) \; =$

Answer: $0 \cdot 1605$ (to 4 significant figures)

(c) $\sqrt{} \; (\; 9 \; . \; 2 \; x^y \; 3 \; -$

$4 \; . \; 5 \; x^2 \;) \; =$

Answer: $27 \cdot 54$ (to 2 decimal places)

(d) $2 \; a^b/c \; 5 \; + \; 3 \; a^b/c \; 4 \; a^b/c \; 7 \; =$

Answer: $3\frac{34}{35}$

ALGEBRAIC MANIPULATION

Addition and subtraction of like terms
- Only add and subtract like terms. Terms like $7x^2$ are different from terms like $5x$.
- Change the order of additions and subtractions provided you move the sign too.

Example Simplify these.
(a) $3a - 4b - a + 7b$ (b) $3x^2 - 2x - 8 + 4x^2 + 7x - 3$

Solution (a) $3a - a - 4b + 7b = 2a + 3b$
(b) $3x^2 + 4x^2 - 2x + 7x - 8 - 3 = 7x^2 + 5x - 11$

Multiplying and dividing terms
When multiplying and dividing treat the numbers and the letters separately.

Example Simplify these.
(a) $5 \times 3a$ (b) $3x \times 2x$ (c) $5a \times 3b$
(d) $3x^2 \times 2x$ (e) $\dfrac{6x^2}{2x}$ (f) $\dfrac{6xy}{3y}$

Solution (a) $15a$ (b) $6x^2$ (c) $15ab$
(d) $6x^3$ (e) $3x$ (f) $2x$

Multiplying brackets
When multiplying out brackets, multiply all terms inside the bracket by the term outside the bracket.

Example Expand these.
(a) $3(x + 5)$ (b) $a(a - 2)$ (c) $4(x - 1) - 2(x - 5)$

Solution (a) $3x + 15$ (b) $a^2 - 2a$ (c) $4x - 4 - 2x + 10 = 2x + 6$

Factorising
When factorising, look for a number or letter (or both) that is a factor of each term.

Example Factorise these.
(a) $5x - 10$ (b) $a^2 + 6a$ (c) $3x^2 + 6xy$

Solution (a) 5 is a factor of each term so $5x - 10 = 5(x - 2)$
(b) a is a factor of each term so $a^2 + 6a = a(a + 6)$
(c) $3x$ is a factor of each term so $3x^2 + 6xy = 3x(x + 2y)$

Practice questions

1 Simplify these.
   (a) $2x + 3 + 5x + 9$ **(b)** $3x - 5y + 2x - 4y$
   (c) $5a - b - 2a + 3b$ **(d)** $3x - 5y - 2 + 4x - 3y + 7$
   (e) $5x^2 + 3x - 2 - x^2 - 7x + 6$ **(f)** $2x^2 - 5x + 2 - 5x^2 - 4x - 7$

2 Simplify these.
   (a) $4a \times 3$ **(b)** $5b \times 2b$ **(c)** $6x \times 3x$
   (d) $5y^2 \times 2y$ **(e)** $6a \times 3b$ **(f)** $2a \times 3b \times 5c$

3 Simplify these.
   (a) $\dfrac{6a}{3a}$ **(b)** $\dfrac{5a^2}{a}$ **(c)** $\dfrac{12x^2}{4x}$

   (d) $\dfrac{15x^3}{3x}$ **(e)** $\dfrac{20xy}{5x}$

4 Expand these.
   (a) $5(a + 4)$ **(b)** $7(x - 2)$ **(c)** $x(x + 3)$
   (d) $a(b - 3)$ **(e)** $3(a + b + c)$

5 Factorise these.
   (a) $2x + 4$ **(b)** $3a - 12$ **(c)** $x^2 + 5x$
   (d) $xy - 3y$ **(e)** $5x + 5y - 5z$

6 Expand and simplify these.
   (a) $3(x + 2) + 5(x + 1)$
   (b) $5(a - 3) + 2(a + 5)$
   (c) $4(x - 3) - 2(x + 3)$

7 Simplify these.
   (a) $\dfrac{x^2 \times x^4}{x^3}$ **(b)** $\dfrac{3a^2 \times 4a^2}{2a}$

8 Factorise these completely.
   (a) $2x^2 + 6x$ **(b)** $5a^2 - 10a$ **(c)** $6x^2 + 3x$
   (d) $4a^2 - 6ab$ **(e)** $5xy + 10yz$

Shorthand in algebra
- $2ac$ means $2 \times a \times c$
- $3t^2$ means $3 \times t^2 = 3 \times t \times t$

Order of operations
- Work out any brackets first
- then powers
- then do multiplying and dividing
- last, add and subtract.

Example Find the value of $y = 2x^2 - 7x$ when $x = 3$.

Solution $y = 2 \times 3^2 - 7 \times 3$
$\quad\quad = 2 \times 9 - 21$
$\quad\quad = 18 - 21$
$\quad\quad = -3$

> **TIP!**
> Work in stages. Write down the answers to the steps.

Substituting negative numbers
- To add and subtract, use a number line.
- To multiply and divide, remember:

  plus $\times$ minus = minus
  minus $\times$ minus = plus
  plus $\div$ minus = minus
  minus $\div$ plus = minus
  minus $\div$ minus = plus

Example Find the value of $\dfrac{ab - c}{2}$ when $a = 3$, $b = -4$ and $c = -6$.

Solution $\dfrac{3 \times -4 - -6}{2} = \dfrac{-12 + 6}{2}$
$\quad\quad\quad\quad = \dfrac{-6}{2}$
$\quad\quad\quad\quad = -3$

> **TIP!**
> Check that you can work with negative numbers using your calculator as well as without it.

Substituting fractions or decimals
For example you may be using a formula you know to calculate an area or volume.

Example $A = \frac{1}{2}(a + b)h$. Find the value of A when $a = 2\cdot4$, $b = 3\cdot8$ and $h = 3\cdot2$.

Solution $A = \frac{1}{2}(2\cdot4 + 3\cdot8) \times 3\cdot2$
$\quad\quad = 3\cdot1 \times 3\cdot2$
$\quad\quad = 9\cdot92$

Practice questions

1 Calculate the value of these expressions when $a = 2$, $b = 3$ and $c = -4$.
(a) $5a$ (b) bc (c) $3a - b$ (d) $5b + c$
(e) abc (f) $ab + 2c$ (g) $6(a + 7)$ (h) $5b^2$
(i) $2c^2$ (j) $a(3b + c)$

2 Find the value of these expressions when $m = 6$, $p = -4$.
(a) $3p$ (b) $p - 2$ (c) $2m + p$
(d) $m - p$ (e) $3(2m + 5)$

3 Find the value of $y = 4x + 1$ when x is
(a) 3   (b) 0   (c) -2

4 Find the value of $y = 3x - 5$ when x is
(a) 6   (b) 1   (c) -2

5 Find the value of $y = x^2 - 2x$ when x is
(a) 4   (b) 1   (c) -3

6 Complete this table for $y = x^2 + 3$.

x	-2	-1	0	1	2	3	4
y	7		3		7	12	

7 The area, A, of a trapezium is given by $A = \frac{1}{2}(a + b)h$.
Find A when $a = 2$, $b = 5$ and $h = 3$.

8 $P = ab + b^2$.
Work out the value of P when   (a) $a = 2$, $b = 3$ (b) $a = 4$, $b = -5$

9 $C = \dfrac{(F - 32) \times 5}{9}$. Find C when $F = 68$.

10 The nth term of a sequence is $7n - 6$. Find the first three terms of this sequence.

11 Complete this table for $y = 8 - 2x$.

x	0	1	2	3	4	5
y		6		2		

12 The area of a triangle is given by $A = \frac{1}{2}bh$. Find A when $b = 6\,\text{cm}$ and $h = 1\frac{1}{2}\,\text{cm}$.

13 Find the value of $2cd$ when $c = 3 \cdot 2$ and $d = 5$.

14 Find the value of $3e - f$ when $e = 4 \cdot 1$ and $f = 5 \cdot 8$.

15 The perimeter of a regular pentagon of side y is given by $P = 5y$. Find P when $y = 3 \cdot 6\,\text{cm}$.

16 The circumference of a circle is given by $C = \pi d$. Find C when $d = 7 \cdot 8\,\text{cm}$.

17 The area of a circle is given by $A = \pi r^2$. Find A when $r = 5 \cdot 2\,\text{cm}$.

18 Find the value of $2a^2$ when $a = 1 \cdot 5$.

19 The surface area of a cube of side x is $6x^2$. Find the surface area of a cube of side $2 \cdot 7\,\text{cm}$.

LINEAR EQUATIONS

> Solve equations by doing the same to each side.

Example   Solve $5x - 3 = 7$.

Solution

$$5x - 3 = 7$$
Add 3 to each side
$$5x = 10$$
Divide each side by 5
$$x = 2$$

Example   Solve $2x + 5 = 12$.

Solution

$$2x + 5 = 12$$
Subtract 5 from each side
$$2x = 7$$
Divide each side by 2
$$x = 3 \cdot 5$$

Example   Solve $\frac{x}{4} = 12$.

Solution

$$\frac{x}{4} = 12$$

Multiply each side by 4
$$x = 48$$

> **TIP!**
> Do not be tempted to divide.

> Solve harder equations using the same methods – collecting terms on each side and then multiplying or dividing to give the answer.

Example   Solve these.

(a) $5x + 17 = 3(x + 6)$ **(b)** $\frac{(15 - x)}{4} = 2$ **(c)** $\frac{17 - x}{4} = 2 - x$

Solution   **(a)**

$$5x + 17 = 3(x + 6)$$
Multiply out brackets
$$5x + 17 = 3x + 18$$
Collect terms
$$5x - 3x = 18 - 17$$
$$2x = 1$$
$$x = \frac{1}{2}$$

(b)

$$\frac{(15 - x)}{4} = 2$$
Multiply both sides by 4
$$15 - x = 8$$
Collect terms
$$7 = x$$

(c)

$$\frac{17 - x}{4} = 2 - x$$
Multiply both sides by 4
$$17 - x = 4(2 - x)$$
Multiply out brackets
$$17 - x = 8 - 4x$$
Collect terms
$$4x - x = 8 - 17$$
$$3x = -9$$
$$x = -3$$

Practice questions

Solve these.

1 $3x + 1 = 16$

2 $4x + 3 = 27$

3 $2x - 3 = 1$

4 $5x - 3 = 1$

5 $3x - 7 = 0$

6 $2x + 5 = 20$

7 $6x - 9 = 27$

8 $x + 6 = 16$

9 $x - 4 = 1$

10 $11x - 10 = 12$

11 $5x + 2 = 27$

12 $7x + 8 = 57$

13 $4x - 11 = -81$

14 $x - 8 = -64$

15 $5x + 25 = 30$

16 $4(x - 3) = 2(3 - x)$

17 $4(x - 1) = 3x + 2$

18 $4(1 - 2x) = 3(2 - 3x)$

19 $2(3x - 1) = 2(x + 1)$

20 $5(2x - 1) = 4(4 - x)$

21 $\dfrac{(2x - 1)}{3} = 1$

22 $\dfrac{(3x - 1)}{5} = 4$

23 $\dfrac{(4x + 4)}{9} = 4$

24 $\dfrac{(3x - 4)}{2} = 3 - x$

25 $\dfrac{(5x + 3)}{3} = 4x + 8$

26 $\dfrac{(2x + 6)}{5} = 3x + 9$

27 $\dfrac{5(x + 3)}{2} = x - 3$

TRIAL AND IMPROVEMENT

- Substitute values into expressions.
- Show the values you try **and** the outcomes.
- The answer to questions like these is always the trial value, not the outcome.

Example A box is a cuboid with a volume of 96 cm³. The length of the box is 2 cm more than its width and the height of the box is 2 cm less than its width. Use trial and improvement to find the dimensions of the box to 1 decimal place.

Solution Start by trying whole numbers. You are trying to find the dimensions to 1 d.p. *not* the volume to 1 d.p.

> **TIP!**
> The answer is **not** the outcome.

Width	Length (w + 2)	Height (w − 2)	Outcome	Too big or too small?
5	7	3	105	Too big, smaller width than 5 needed
4	6	2	48	Too small, larger width than 4 needed
				Try something between 4 and 5
4·5	6·5	2·5	73·125	Too small, larger width than 4·5 needed
4·7	6·7	2·7	85·023	Too small, larger width than 4·7 needed
4·8	6·8	2·8	91·392	Too small, larger width than 4·8 needed
4·9	6·9	2·9	98·049	Too big, smaller width than 4·9 needed
				Try something between 4·8 and 4·9
4·85	6·85	2·85	94·684 125	Too small, width is larger than 4·85 which, to 1 decimal place, is 4·9 cm

Width = 4·9 cm, length = 6·9 cm and height = 2·9 cm, each to 1 decimal place.

Example One solution to $x^3 - 3x = 13$ lies between 2 and 3. Use trial and improvement to find it. Give your answer to 2 decimal places.

Solution

Value	x^3	$-3x$	Outcome	Too big or too small?
2	8	−6	2	Too small, larger value than 2 needed
3	27	−9	18	Too large, smaller value than 3 needed
				Try something between 2 and 3
2·5	15·625	−7·5	8·125	Too small, larger value than 2·5 needed
2·75	20·796 875	−8·25	12·546 875	Too small, larger value than 2·75 needed
2·8	21·952	−8·4	13·552	Too large, smaller value than 2·8 needed
				Try something between 2·75 and 2·8
2·77	21·253 933	−8·31	12·943 933	Too small, larger value than 2·77 needed
2·78	21·484 952	−8·34	13·144 952	Too large, smaller value than 2·78 needed
2·775	21·369 234	−8·325	13·044 234	Too large, answer is smaller than 2·775 which, to 2 decimal places, is 2·77

$x = 2·77$ to 2 decimal places.

Practice questions

1 When three numbers x, $x + 1$ and $x + 3$ are multiplied together the answer is 123. Use trial and improvement to find the three numbers, each to 1 decimal place.

2 A box is a cuboid with a volume of $500\,cm^3$. The length of the box is 10 cm more than its width and the height of the box is 5 cm less than its width. Use trial and improvement to find the dimensions of the box to 1 decimal place.

3 A box is a square-based cuboid with a volume of $127\,cm^3$. The height of the box is 3 cm less than its length. Use trial and improvement to find the dimensions of the box, each to 2 decimal places.

4 A box is a cuboid with a volume of $1000\,cm^3$. The width of the box is double its height and the length of the box is three times its height. Use trial and improvement to find the dimensions of the box, each to 1 decimal place.

5 One solution to $x^3 = 17$ lies between 2 and 3. Use trial and improvement to find it. Give your answer to 1 decimal place.

6 One solution to $x^3 - x^2 = 3$ lies between 1 and 2. Use trial and improvement to find it. Give your answer to 1 decimal place.

7 One solution to $x^3 + 19 = 0$ lies between -2 and -3. Use trial and improvement to find it. Give your answer to 1 decimal place.

8 One solution to $x^3 + 3x = 50$ lies between 3 and 4. Use trial and improvement to find it. Give your answer to 2 decimal places.

9 One solution to $x^3 - 3x^2 - 5 = 0$ lies between 3 and 4. Use trial and improvement to find it. Give your answer to 2 decimal places.

10 One solution to $x^3 + 3x = 31$ lies between 2 and 3. Use trial and improvement to find it. Give your answer to 2 decimal places.

DRAWING GRAPHS

> To draw a graph from its equation, first
> use the equation to work out the values
> of y for suitable values of x. Plot each
> point (x, y) and join the points.

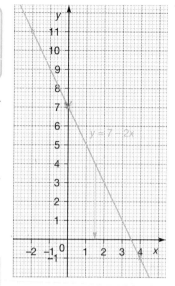

Example (a) Draw the graph of $y = 7 - 2x$
for $x = -2$ to 4.
(b) From the graph, find the
value of x when $y = 4$.

Solution (a)

x	-2	0	4
y	11	7	-1

(b) When $y = 4$, $x = 1.5$.

TIP!
Work out at least three points.
Watch the signs.

TIP!
Find where $y = 4$ on the graph
and read the value of x.

Example (a) Copy and complete the table for $y = 2x^2 - 3x - 6$.

x	-2	-1	0	1	2	3
y		-1	-6		-4	3

TIP!
$2x^2$ is $2 \times x \times x$

(b) Draw the graph of $y = 2x^2 - 3x - 6$ for $x = -2$ to 3.
(c) Write down the values of x when $y = 0$.

Solution (a)

x	-2	-1	0	1	2	3
y	8	-1	-6	-7	-4	3

(b)

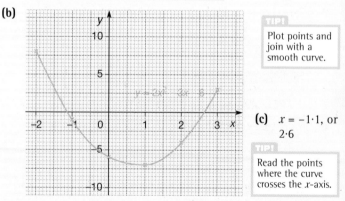

TIP!
Plot points and
join with a
smooth curve.

(c) $x = -1.1$, or
2.6

TIP!
Read the points
where the curve
crosses the x-axis.

Practice questions

1 **(a)** Copy and complete this table for the equation $y = 3x + 2$.

x	−3	−2	−1	0	1	2
y		−4				

(b) Draw the graph of $y = 3x + 2$ for $x = -3$ to 2.

2 **(a)** Copy and complete this table for the equation $y = 2 - x$.

x	−2	−1	0	1	2	3
y	4	3	2	1	0	−1

(b) Draw the graph of $y = 2 - x$ for $x = -2$ to 3.

3 **(a)** Copy and complete this table for the equation $2y = x + 4$.

x	−4	−3	−2	−1	0	1	2
y	0	1	1	2	2	2·5	3

(b) Draw the graph of $2y = x + 4$ for $x = -4$ to 2.

4 **(a)** Draw the graph of $y = 2x - 1$ for $x = -2$ to 4.
   (b) From the graph, find the value of x when $y = 4$.

5 **(a)** Draw the graph of $y = 2x + 3$ for $x = -2$ to 4.
   (b) On the same graph, draw the graph of $2y = 8 - x$ for $x = -2$ to 4.
   (c) Write down the coordinates of the point where the two lines cross.

6 **(a)** Copy and complete this table for $y = x^2 - 6$.

x	−3	−2	−1	0	1	2	3
y	3			−6	−5	−2	3

(b) Draw the graph of $y = x^2 - 6$ for $x = -3$ to +3.
   (c) Write down the values of x where $y = 0$.

7 **(a)** Copy and complete this table for $y = x^2 - x - 6$.

x	−3	−2	−1	0	1	2	3	4
y	6		−4		−6			

(b) Draw the graph of $y = x^2 - x - 6$ for $x = -3$ to +4.
   (c) **(i)** On the same axes, draw the graph of $y = x$.
       (ii) Write down the coordinates of the points where they cross.

8 Copy and complete this table for $y = 2x^2 + 2x - 5$.

x	−3	−2	−1	0	1	2
y	7	−1		−5		

(b) Draw the graph of $y = 2x^2 + 2x - 5$ for $x = -3$ to +2.
   (c) Write down the values of x where $y = 0$.

Angle facts
- Angles on a straight line add to 180°.
- Angles around a point add to 360°.
- Angles in a triangle add to 180°.
- Vertically opposite angles are equal.
- In an isosceles triangle two angles are equal.
- Angles in a quadrilateral add to 360°.

Example Work out these angles. Give a reason for each answer.

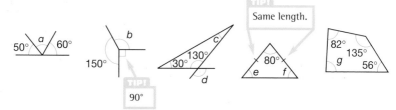

TIP!
Same length.

TIP!
90°

Solution
$a = 180 - (50 + 60) = 70°$, angles on a straight line
$b = 360 - (150 + 90) = 120°$, angles around a point
$c = 180 - (130 + 30) = 20°$, angles in a triangle
$d = 130°$, vertically opposite angles
$e = f = (180 - 80) \div 2 = 50°$, isosceles triangle
$g = 360 - (56 + 82 + 135) = 87°$, angles in a quadrilateral

Angles with parallel lines
- Alternate (Z) angles are equal.
- Corresponding (F) angles are equal.
- Allied (U or C) angles add to 180°.

Example Work out these angles. Give a reason for each answer.

(a)

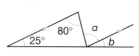

(b)

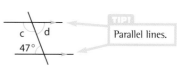

TIP!
Parallel lines.

Solution (a) $a = 80°$, alternate angles $b = 25°$, corresponding angles

(b) $c = 180 - 47 = 133°$, allied angles $d = 47°$, alternate angles

Practice questions

Work out the angles marked with letters in these diagrams.
Give a reason for each answer.

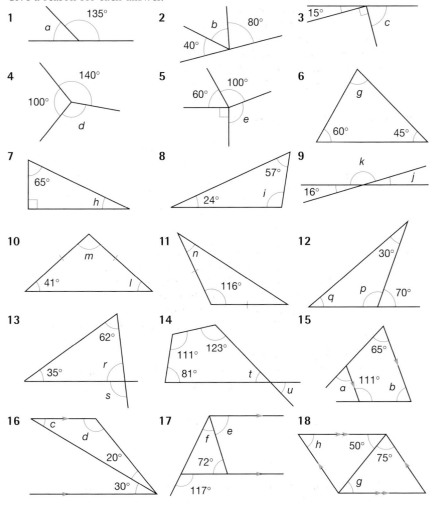

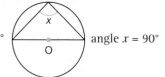

- The angle in a semi-circle = 90°   angle x = 90°

- The angle at the centre of a circle = 2 × angle at the circumference.

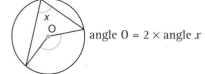

  angle O = 2 × angle x

- Angles in the same segment of a circle are equal, so angles made by

  the same arc are equal.   angle x = angle y

Example O is the centre of the circle.
Calculate p, q, r and s.

Solution $p = 90° - 55° = 35°$
(angle in a semi-circle)

$q = 35°$ (triangle OBC is isosceles)

$r = 180° - 35° - 35° = 110°$
(angle sum of triangle)

$s = 110 \div 2 = 55°$ (angle at circumference)

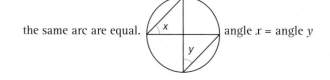

Example Find the size of angles x and y.
Give a reason for each answer.

Solution $x = 90°$, angle between tangent and radius
$y = 180 - (90 + 37) = 53°$, angles in a triangle

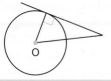

Tangents to a circle
The angle between a tangent and the radius is 90°.

Practice questions

1 O is the centre of each circle. Calculate the lettered angles.

(a)

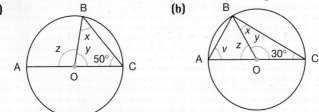

(b)

2 O is the centre of each circle. Calculate the lettered angles.

(a)

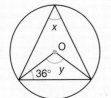

(b)

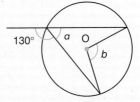

3 O is the centre of each circle. Calculate the lettered angles.

(a)

(b)

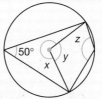

4 O is the centre of each circle. Calculate the lettered angles.

(a)

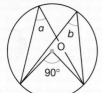

(b)

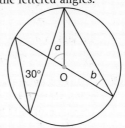

5

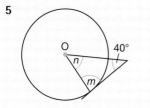

6

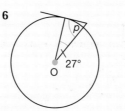

7

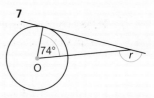

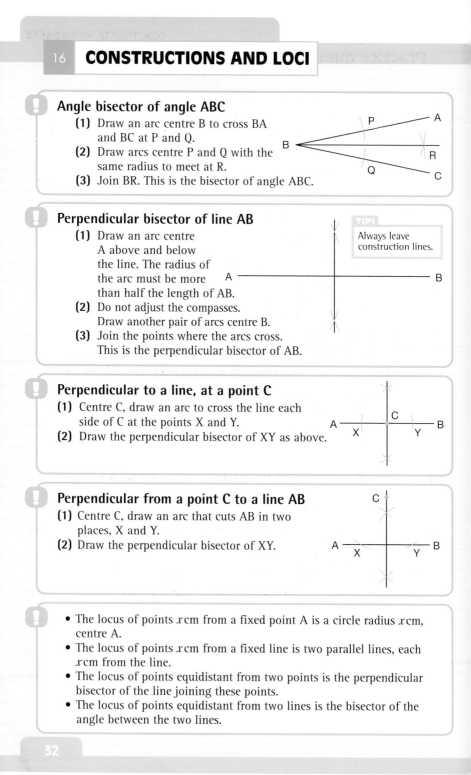

Angle bisector of angle ABC

(1) Draw an arc centre B to cross BA and BC at P and Q.

(2) Draw arcs centre P and Q with the same radius to meet at R.

(3) Join BR. This is the bisector of angle ABC.

Perpendicular bisector of line AB

(1) Draw an arc centre A above and below the line. The radius of the arc must be more than half the length of AB.

(2) Do not adjust the compasses. Draw another pair of arcs centre B.

(3) Join the points where the arcs cross. This is the perpendicular bisector of AB.

TIP! Always leave construction lines.

Perpendicular to a line, at a point C

(1) Centre C, draw an arc to cross the line each side of C at the points X and Y.

(2) Draw the perpendicular bisector of XY as above.

Perpendicular from a point C to a line AB

(1) Centre C, draw an arc that cuts AB in two places, X and Y.

(2) Draw the perpendicular bisector of XY.

- The locus of points x cm from a fixed point A is a circle radius x cm, centre A.
- The locus of points x cm from a fixed line is two parallel lines, each x cm from the line.
- The locus of points equidistant from two points is the perpendicular bisector of the line joining these points.
- The locus of points equidistant from two lines is the bisector of the angle between the two lines.

Practice questions

In all constructions, use compasses and leave the construction lines in.

1 Draw these angles, using a protractor or angle measurer.
Construct the angle bisectors and check your accuracy with a protractor.
(a) 60°   **(b)** 35°   **(c)** 130°

2 Draw lines with these lengths.
Construct the perpendicular bisectors and check your accuracy with a ruler and a protractor.
(a) 10 cm   **(b)** 5 cm   **(c)** 11·5 cm

3 Draw a line about 12 cm long.
Mark a point on the line 5 cm from one end.
Construct the perpendicular to the line at that point.

4 Draw a line about 10 cm long.
Mark a point, not on the line, about 5 cm away from the line.
Construct the perpendicular to the line from the point.

5 Draw the locus of the points that are 3 cm from a fixed point A.

6 Draw the triangle ABC with AB = 5 cm, AC = 4 cm and angle A = 60°.
Construct the locus of the points that are equidistant from AB and AC.

7 Draw the triangle ABC with AB = 6 cm, AC = 5 cm and BC = 3·5 cm.
Construct the locus of the points that are equidistant from A and B.

8 Draw the triangle ABC with AB = 6 cm, AC = 5 cm and BC = 7 cm. Bisect the angle A. Shade the locus of the points inside the triangle and nearer to AB than AC.

9 Draw the triangle ABC with AB = 6 cm, AC = 5 cm and A = 40°. Draw appropriate constructions and show the region inside the triangle, that is nearer to A than B and nearer to BC than AB.

10 Three towns, Dinchester (D), Bycastle (B) and Sevenhills (C) are the following distances apart. DC = 20 km, DB = 15 km, BC = 25 km. BC is East West and D is North of the other two.
(a) Make a scale drawing of the three places, use a scale of 2 cm to 5 km.

It is proposed that an outlet is to be built. It must be nearer to B than C and not more than 10 km from D.

(b) Show by shading the region where the outlet can be built. Label it R.

TRANSFORMATIONS

Reflection

The image is the opposite side of the mirror line, the same distance away from it as the object.

Example Reflect the triangle in the line $y = 2$.

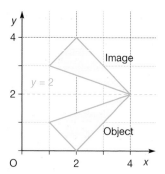

Enlargement

- The length of each side of the object is multiplied by the scale factor.
- The distance of the image from the centre of enlargement is the scale factor × the distance of the object from the centre of enlargement.

Example Enlarge the flag with centre (0, 0) and scale factor 2.

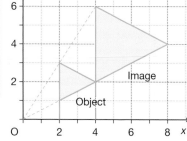

Rotation

- Use tracing paper to find the correct position. Use a pencil to keep the centre of rotation still. Rotate through the correct angle – and also in the correct direction, clockwise or anticlockwise.

Translation

- Every point on the object moves the same distance and direction.

TIP!
Don't confuse the words transformation and translation.

Describing transformations

- Make sure you give all the information needed.
  For example, for a rotation give the centre of rotation, the angle and whether it is clockwise or anticlockwise.
- A translation may be described using a column vector.
  For example, $\begin{pmatrix} 3 \\ -4 \end{pmatrix}$ means 3 units to the right and 4 down.

Practice questions

1 On squared or graph paper, draw axes with x and y from –6 to 6. Draw triangle P with vertices at (1, 1) (2, 1) and (1, 3).
  (a) Reflect P in the x-axis. Label the image A.
  (b) Reflect P in the line $y = 3$. Label the image B.
  (c) Rotate P about the origin through 90° anticlockwise. Label the image C.
  (d) Rotate P through 180° about the point (–1, –1). Label the image D.

2 On squared or graph paper, draw axes with x and y from –6 to 6. Draw triangle P with vertices as (1, 1) (2, 1) and (1, 3).
  (a) Enlarge P with centre (0, 0) and scale factor 2. Label the image E.
  (b) Enlarge P with centre (3, 2) and scale factor 3. Label the image F.
  (c) Translate P 2 units to the right and 5 down. Label the image G.

3

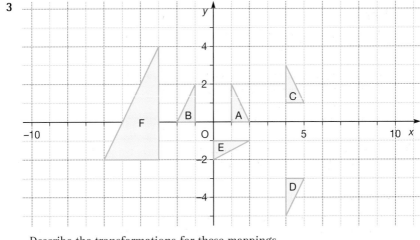

Describe the transformations for these mappings.
  (a) A onto B   **(b)** A onto C   **(c)** C onto A
  (d) C onto D   **(e)** A onto E   **(f)** B onto E

4 Using the diagram to question 3, describe the transformations for these mappings.
  (a) B onto F   **(b)** F onto B   **(c)** B onto D

5 On squared or graph paper, draw axes with x and y from –6 to 6. Draw triangle Q with vertices at (4, 2) (6, 2) and (4, 6).
  (a) Enlarge triangle Q with centre (0, 0) and scale factor $\frac{1}{2}$. Label the image J.
  (b) Rotate J through 90° clockwise about (0, –2). Label the image K.
  (c) Reflect Q in the line $y = -x$. Label the image L.
  (d) Translate Q by $\begin{pmatrix} -3 \\ -8 \end{pmatrix}$. Label the image M.

PYTHAGORAS

> • In a right-angled triangle. the longest side (the hypotenuse) is opposite the right angle.
> • Pythagoras' theorem states 'The square on the hypotenuse is equal to the sum of the squares on the other two sides'.
> • When two sides are given in a right-angled triangle, use Pythagoras to find the third side.

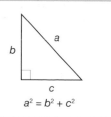

$$a^2 = b^2 + c^2$$

Example Find the third side in each of these triangles. All lengths are in centimetres.

(a)   3·7   (b)

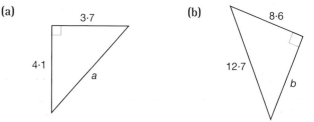

Solution (a) The longest side (the hypotenuse) is needed.
So $a^2 = 4·1^2 + 3·7^2 = 16·81 + 13·69 = 30·5$
$a = \sqrt{30·5} = 5·52$ cm (to 2 decimal places).

(b) One of the shorter sides is needed.
So $b^2 = 12·7^2 - 8·6^2 = 161·29 - 73·96 = 87·33$
$b = \sqrt{87·33} = 9·35$ cm (to 2 decimal places).

Example A rectangle has sides 5·6 cm and 4·8 cm. Find the length of the diagonal.

Solution

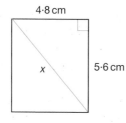

The diagonal is the hypotenuse of the right-angled triangle.
So $x^2 = 4·8^2 + 5·6^2 = 23·04 + 31·36 = 54·4$
$x = \sqrt{54·4} = 7·38$ cm
(to 2 decimal places).

Practice questions

In the following six questions find the length of the side marked with a letter. Give each answer exactly or to 1 decimal place.

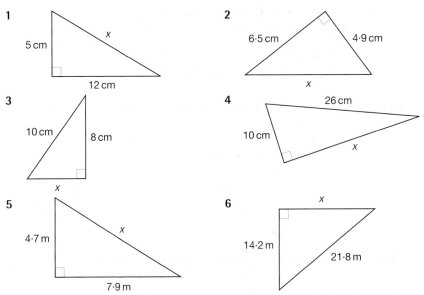

1 5 cm x 12 cm

2 6·5 cm 4·9 cm x

3 10 cm 8 cm x

4 26 cm 10 cm x

5 4·7 m x 7·9 m

6 x 14·2 m 21·8 m

7 A right-angled triangle has the two shorter sides 9 cm and 12 cm long. Find the length of the hypotenuse.

8 The hypotenuse of a right-angled triangle is 14 cm and one of the shorter sides is 8 cm. Find the length of the other side.

9 An isosceles triangle has a base length of 6 cm and a vertical height of 5 cm. Find the length of the sloping sides.

10 A ladder 4 metres long is on horizontal ground leaning against a wall. The bottom of the ladder is 1·1 m from the wall. The wall is perpendicular to the ground. How far up the wall does the ladder reach?

11 A rectangle has one side 4·5 cm long and the diagonal is 7·5 cm long. Find the length of the other side.

12 There is a path round two sides of a rectangular lawn. The two lengths are 16 m and 26 m. It is decided to make a path diagonally across the lawn. How long will the path be?

13 On a grid a line joins the origin to the point (4, 6). Find the length of the line.

14 Two poles are fixed 6 metres apart perpendicular to horizontal ground. They are 2 m and 3·5 m high. How far is it between their top points?

15 A boat is at sea 50 m from the bottom of a vertical cliff 35 m high. How far is it from the boat to the top of the cliff?

AREA, VOLUME AND PERIMETER

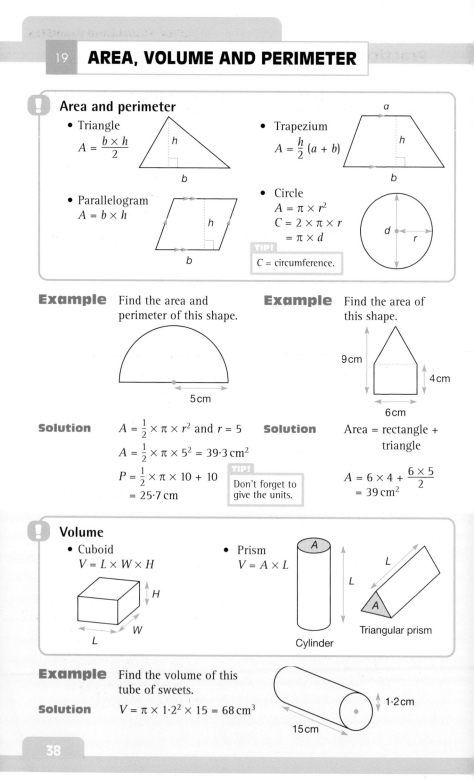

Area and perimeter

- Triangle
  $$A = \frac{b \times h}{2}$$

- Parallelogram
  $$A = b \times h$$

- Trapezium
  $$A = \frac{h}{2}(a + b)$$

- Circle
  $$A = \pi \times r^2$$
  $$C = 2 \times \pi \times r$$
  $$= \pi \times d$$

TIP!
C = circumference.

Example Find the area and perimeter of this shape.

Solution
$A = \frac{1}{2} \times \pi \times r^2$ and $r = 5$

$A = \frac{1}{2} \times \pi \times 5^2 = 39 \cdot 3 \text{ cm}^2$

$P = \frac{1}{2} \times \pi \times 10 + 10$

$= 25 \cdot 7 \text{ cm}$

TIP!
Don't forget to give the units.

Example Find the area of this shape.

9 cm

4 cm

6 cm

Solution Area = rectangle + triangle

$A = 6 \times 4 + \frac{6 \times 5}{2}$

$= 39 \text{ cm}^2$

Volume

- Cuboid
  $$V = L \times W \times H$$

- Prism
  $$V = A \times L$$

Cylinder

Triangular prism

Example Find the volume of this tube of sweets.

Solution $V = \pi \times 1 \cdot 2^2 \times 15 = 68 \text{ cm}^3$

1·2 cm

15 cm

Practice questions

1 A rectangle has a length of 10 cm and a perimeter of 32 cm.

Find the width and the area of this rectangle.

2 Find the area and the perimeter of this triangle.

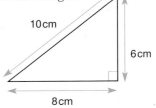

10 cm

6 cm

8 cm

3 Find the area of this parallelogram.

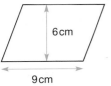

6 cm

9 cm

4 Find the area of this trapezium.

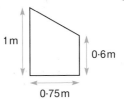

1 m

0·6 m

0·75 m

5 Find the area of this shape.

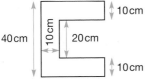

35 cm

10 cm

40 cm

10 cm

20 cm

10 cm

6 Find the area of this kite.

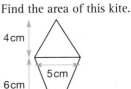

4 cm

5 cm

6 cm

7 Find the volume and the surface area of this cuboid.

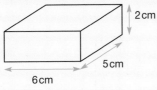

2 cm

5 cm

6 cm

8 Find the volume and the surface area of this triangular prism.

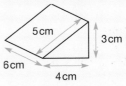

5 cm

3 cm

6 cm

4 cm

9 Find the volume of this semi-circular tent.

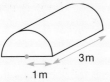

3 m

1 m

10 The volume of water in this tank is 3000 cm³. Find the depth of the water.

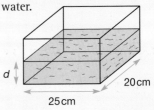

d

20 cm

25 cm

Speed, distance, time

- Relationships connecting speed, distance and time are summed up in this triangle.

$$\text{speed} = \frac{\text{distance}}{\text{time}} \qquad \text{time} = \frac{\text{distance}}{\text{speed}} \qquad \text{distance} = \text{speed} \times \text{time}$$

- Some units of speed are miles per hour (m.p.h.), kilometres per hour (km/h) and metres per second (ms^{-1}).

Example Change 2 hours 53 minutes to hours .

Solution $53 \div 60 = 0\cdot883$ So 2 hours 53 minutes = $2\cdot883$ hours

Example A train travels 200 miles in 3 hours. What is its average speed?

Solution Average speed = $\frac{200}{3}$ = $66\cdot7$ m.p.h.

Example A car travels at 95 km/h for 2 hours 53 minutes. How far does it travel?

Solution As above, 2 hours 53 minutes = $2\cdot883$ hours
Distance = $95 \times 2\cdot883 = 274$ km

Example This graph shows Mark's cycle journey.

(a) How far from his home did he get?
(b) Calculate his greatest speed.
(c) On section EF the line slopes down. What does this show?

Solution (a) 10 km
(b) Section AB (the steepest section), speed = $6 \div 0\cdot5 = 12$ km/h
(c) That Mark was on his way back home.

Other compound units

Other compound units are:
- density in grams per cubic centimetre (g/cm^3)
- petrol consumption in miles per gallon (m.p.g.) or kilometres per litre (km/l)
- population density in people per square kilometre (people/km^2)

Example The population of Wales at the 1991 census was 2 835 073. If the area of Wales is 20 758 km^2, what was the population density in 1991?

Solution Population density = $\frac{2\,835\,073}{20\,758}$ = 137 people/km^2

Practice questions

1 The graph on the right shows a coach
journey from Nottingham to Guildford.
(a) How far did the coach travel?
(b) What does the horizontal line on
the graph show?
(c) What was the average speed
during the last part of the
journey?

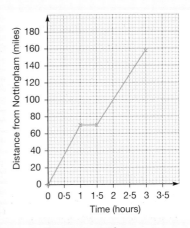

2 Draw a graph for Jenny's journey from home to Newport, 120 km away.
She travels 80 km in 1·5 hours, stops for 1 hour and does the rest in $1\frac{1}{4}$ hours.
(a) Calculate Jenny's speed on the last stage of her journey.

At the same time as Jenny leaves home, Tim leaves Newport and travels
towards Jenny's home. He travels at 60 km/h.
(b) Draw a line to show Tim's journey and find where they pass each other.

3 A car travels 180 miles in 4 hours. What is the average speed?

4 In the Le Mans 24-hour race the winning team's average speed was 217 km/h.
How far did they travel?

5 A cyclist travelled 30 miles at 12 m.p.h. How long did it take?

6 A train travelled 320 km in 2 hours 15 minutes. Find the average speed.

7 Sarah walked 3 km in 45 minutes. Mike walked 4·8 km in 1 hour 15 minutes.
Who walked the faster? Show your calculations.

8 A runner ran for 50 s at $8·2 \, \text{ms}^{-1}$. How far did he run?

9 On a sponsored walk, Lesley walked 36 km at 3·4 km/h. How long did it take her?
Give your answer in hours and minutes.

10 In the 1991 census the population of Scotland was 4 998 567. If the area of
Scotland is $78\,789 \, \text{km}^2$, what was the population density in 1991?

11 A car travelled 450 miles and used 11 gallons of petrol. What was the
average petrol consumption in miles per gallon?

12 A cube of steel of side 10 cm has a mass of 8500 g. What is the density of
steel in g/cm^3?

SCATTER DIAGRAMS

> Scatter diagrams show points of data on a grid and the points are not joined.

Example Plot these points showing height and spending money for some children.

Height (cm)	150	130	165	135	140
Money (£)	4·00	2·00	3·50	2·50	3·50

Height (cm)	170	155	160	145	155
Money (£)	4·50	3·50	3·00	2·50	5·00

Solution

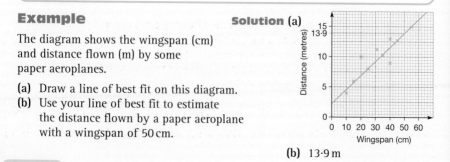

> ### Correlation
> • The correlation (relationship), if any, between two sets of data is described by direction (positive or negative) and strength (strong or weak).

Example Describe the correlation shown in each of these diagrams.

(a)   (b)   (c)   (d)

TIP!
The first example (above) shows weak positive correlation.

Solution   (a) Strong positive   (b) Weak negative
               (c) Strong negative   (d) No correlation

> ### Line of best fit
> A line of best fit is a straight line passing as closely as possible to the points.

Example

The diagram shows the wingspan (cm) and distance flown (m) by some paper aeroplanes.

(a) Draw a line of best fit on this diagram.
(b) Use your line of best fit to estimate the distance flown by a paper aeroplane with a wingspan of 50 cm.

Solution (a)

(b) 13·9 m

Practice questions

1   The table below shows data about the number of cars and the number of
    people killed on the road. Draw a scatter graph to show the data.

Cars (million)	20·2	20·3	20·7	20·8	21·7	22·2	22·8	23·1	23·9
Road deaths (nearest 10)	770	750	760	630	590	580	540	520	480

2   The table below shows data about the length of a person's leg and the time
    it takes them to run 400 metres. Draw a scatter graph to show the data.

Leg length (cm)	75	82	90	77	85	87	79	83	88	82	80	78
Time (seconds)	53·2	51·8	50·9	57·4	53·9	57·5	53.7	52·6	49·4	51·6	56·1	54·3

3   Which of the diagrams below could show the correlation between these?
    (a) Height and arm span
    (b) Amount of sunshine and umbrella sales
    (c) Age and intelligence of teenagers

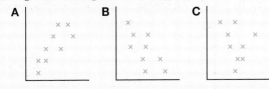

4   This graph shows the leg length and the time
    taken to run 100 m by a group of students.
    (a) Draw a line of best fit on the graph.
    (b) Use your line to estimate the time it would
    take a person with a leg length of 80 cm
    to run 100 metres.

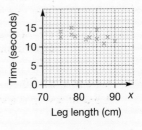

5   The table below shows the attendance figures, and number of arrests, at
    football league grounds over a period of 10 years.
    (a) Draw a scatter graph to show this data.
    (b) Describe the correlation between attendance and arrests.
    (c) Draw a line of best fit for the data.

Attendance (millions)	27·8	26·0	25·4	25·3	24·7	22·8	21·8	21·9	21·7	20·6
Arrests (to nearest 50)	3200	3400	3150	3350	3300	3600	3940	3850	4250	4600

22 CUMULATIVE FREQUENCY AND BOX PLOTS

> A cumulative frequency curve is a way of representing continuous data.

Example This table summarises the heights of a group of 40 people.

Height (h inches)	$60 \leqslant h < 63$	$63 \leqslant h < 67$	$67 \leqslant h < 71$	$71 \leqslant h < 75$	$75 \leqslant h < 79$
Frequency	5	14	14	5	2

The cumulative frequency table for these data is

Height (h inches)	$h < 63$	$h < 67$	$h < 71$	$h < 75$	$h < 79$
Cumulative frequency	5	19	33	38	40

Draw the cumulative frequency curve and estimate the median height.

Solution

There are 40 people so the 'median person' is midway between the 20th and the 21st person.

Find 20·5 on the frequency axis and read off the height.

The median height = 67·4 inches.

TIP!
For a large frequency it is sufficiently accurate to read at half the total for the median and at a quarter and three quarters of the total for the quartiles.

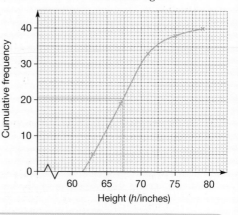

> Box plots can be used to show the spread of a set of data. They show the extreme values, the lower and upper quartiles and the median.

Example This list shows the numbers of CD players sold by a shop over an eleven-day period.
0   2   5   2   0   4   4   8   9   8   8
Show this data on a box plot.

Solution Putting the numbers in order gives
0   0   2   2   4   4   5   8   8   8   9
The median is 4.
The lower quartile is 2.
The upper quartile is 8.

Number sold

Practice questions

1   This cumulative curve is based on the
    heights of the 250 girls in Year 11 at
    Egton Girls School.

    Find these estimates.
    (a) How many of the girls have
    heights of 170 cm or less?
    (b) How many of the girls have
    heights between 160 cm and
    170 cm?
    (c) The median height

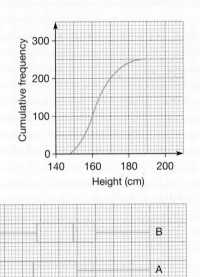

2   These box plots compare the
    age distribution in two towns
    A and B.

    Compare the two distributions.

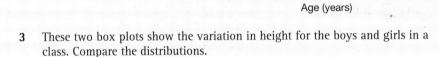

3   These two box plots show the variation in height for the boys and girls in a
    class. Compare the distributions.

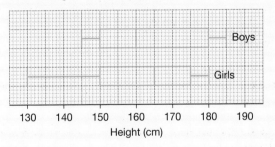

4   Rachael works for five days a week in London. She keeps a record of the
    time, in minutes, it takes her to get to work each day for a month.

    43   56   50   45   51   58   46   61   75   48
    43   49   45   47   54   46   48   53   62   79

    (a) Write these times in order.
    (b) Find the median, the lower quartile and the upper quartile.

FREQUENCY DIAGRAMS AND POLYGONS

- A frequency diagram has bars stretching over the full interval.
- A frequency polygon has straight lines joining midpoints of the intervals.

Example

This is the time spent by 100 students on a computer during one evening.

(a) Draw a frequency diagram of these data.

(b) What is the modal group?

Time (t minutes)	Frequency
$0 \leqslant t < 30$	2
$30 \leqslant t < 60$	5
$60 \leqslant t < 90$	8
$90 \leqslant t < 120$	27
$120 \leqslant t < 150$	34
$150 \leqslant t < 180$	24

Solution

(a)

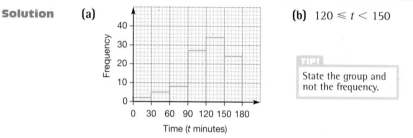

(b) $120 \leqslant t < 150$

TIP!
State the group and not the frequency.

Example

This table gives the hand spans of 20 boys and 20 girls in year 10.

(a) Draw the frequency polygon for the boys and girls on the same graph.

(b) (i) State the boys' modal span.

(ii) State two differences between the boys' and girls' spans.

Span (s cm)	Frequency	
	Boys	Girls
$15{\cdot}0 \leqslant s < 15{\cdot}5$	0	1
$15{\cdot}5 \leqslant s < 16{\cdot}0$	0	3
$16{\cdot}0 \leqslant s < 16{\cdot}5$	2	8
$16{\cdot}5 \leqslant s < 17{\cdot}0$	3	6
$17{\cdot}0 \leqslant s < 17{\cdot}5$	5	2
$17{\cdot}5 \leqslant s < 18{\cdot}0$	4	0
$18{\cdot}0 \leqslant s < 18{\cdot}5$	2	0
$18{\cdot}5 \leqslant s < 19{\cdot}0$	4	0

Solution

(a)

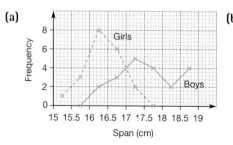

(b) (i) $17{\cdot}0 \leqslant t < 17{\cdot}5$

(ii) The girls have smaller spans (lower modal group).
The boys' spans are more varied than the girls'.

Practice questions

1   The table gives the hours that 200 light bulbs were lit before burning out.

Hours lit (t)	Frequency
$0 \leqslant t < 400$	30
$400 \leqslant t < 800$	50
$800 \leqslant t < 1200$	100
$1200 \leqslant t < 1600$	14
$1600 \leqslant t < 2000$	6

(a) Draw a frequency diagram to illustrate this information.
(b) State the modal class.

2   These are the speeds of 100 cars on a motorway in miles per hour.

Speed	Number of cars
$50 \leqslant s < 55$	2
$55 \leqslant s < 60$	8
$60 \leqslant s < 65$	12
$65 \leqslant s < 70$	37
$70 \leqslant s < 75$	26
$75 \leqslant s < 80$	10
$80 \leqslant s < 85$	5

(a) Draw a frequency polygon to illustrate this information.
(b) State the modal speed.

3   These are the marks of 1000 candidates in two Chemistry exams.

Marks	Paper 1	Paper 2
$0 \leqslant m < 20$	15	10
$20 \leqslant m < 40$	180	95
$40 \leqslant m < 60$	460	480
$60 \leqslant m < 80$	280	335
$80 \leqslant m < 100$	65	80

(a) Draw a frequency polygon for both papers on the same graph.
(b) **(i)** State the modal group for each paper.
      (ii) Make two comments to compare the performances on the two papers.

STEM-AND-LEAF DIAGRAMS

Drawing a stem–and–leaf diagram
- Stems are written to the left of the vertical line and the leaves to the right.
- Leaves for each stem are ordered.
- Each stem-and-leaf diagram must have a key.

Example   A plumber records the time, in hours, it takes him to complete his jobs during one week.

0·9   1·1   2·1   4·2   0·7   2·3   1·6   2·1   0·3   0·8
1·1   0·9   1·8   0·9   1·2   2·7   0·4

Draw a stem-and-leaf diagram for this data.

Solution

```
0 | 9  7  9  9  3  8  4
1 | 1  1  8  2  6
2 | 1  3  1  7
3 |
4 | 2
```

> **TIP!**
> Work through the list systematically. Don't worry about the order.

Ordered

```
0 | 3  4  7  8  9  9  9
1 | 1  1  2  6  8
2 | 1  1  3  7
3 |
4 | 2
```

> **TIP!**
> Now put the leaves in order, smallest nearest the stem.

Key : 4 | 2 means 4·2 hours

Interpreting a stem-and-leaf diagram
These can be found from the stem-and-leaf diagram:
- the number of data items
- the range
- the mode
- the median
- the modal group (the stem which has most leaves).

Example   A police speed camera measured the speed of cars on a road through a village. The data collected is shown in this stem-and-leaf diagram.

```
1 | 2  4  5  6  8
2 | 1  5  5  5  7  8  9
3 | 0  2  4  6  7
4 | 1  3
5 | 2
```
Key : 5 | 2 means 52 m.p.h.

(a) Find the range of the speeds.
(b) Find the median speed.
(c) The speed limit on the road is 40 m.p.h. What percentage of cars were travelling faster than the speed limit?

Solution   (a) $52 - 12 = 40$ m.p.h.   (b) $\dfrac{(27 + 28)}{2} = 27 \cdot 5$ m.p.h.

(c) $\dfrac{3}{20} \times 100 = 15\%$

Practice questions

1 The ages, in years, of people entering a doctor's surgery are listed below.
   Draw a stem-and-leaf diagram for these data.

58	30	81	51	58	91	71	67	56	74
75	48	63	79	57	74	50	34	62	67
75									

2 The weights, in pounds, of 20 students are listed below.
   Draw a stem-and-leaf diagram for these data.
   Use stems from 10 to 17.

118	147	146	138	175	118	155	146	135	127
136	122	114	140	106	159	127	143	153	139

3 The stem-and-leaf diagrams below show the scores of a group of students in
   the two papers of an English exam.

   Paper 1

   ```
   1 | 3 5
   2 | 0 5 7
   3 | 0 4 6 9
   4 | 2 5 7
   5 | 1 5 5 6 8
   6 | 0 4 4 4 8 9
   7 | 0 3 3 6 7 7
   8 | 1 3 4 5
   ```

   Paper 2

   ```
   1 | 0 1 4 7
   2 | 1 1 4 5 7 8
   3 | 2 5 5 5 8 8 9
   4 | 0 2 2 6 7
   5 | 1 3 4 9 9
   6 | 0 2 5
   7 | 3 9
   8 | 1
   ```

   Key: 8 | 1 means 81

   (a) How many students are in the group?
   (b) Find the mode for each paper.
   (c) Find the median mark for each paper.
   (d) Find the range of marks for each paper.
   (e) Which paper was harder? Explain how you know this.

4 The heights, in centimetres, of 30 women are recorded below.

174	160	141	153	161	159	163	186	179	167
154	145	156	159	171	156	142	169	160	171
188	151	162	164	172	181	152	178	151	177

   (a) Draw a stem-and-leaf diagram for these data.
   (b) What is the modal group?
   (c) What percentage of the women has a height greater than 170 cm?
   (d) One woman is chosen from the group at random.
       What is the probability that she is under 150 cm tall?

> ● The whole pie (360°) represents the total frequency.
> ● $\dfrac{360°}{\text{total frequency}}$ represents each item.

Example The table shows how the 30 members of Form 11H get to school.

Method of travel	Bus	Car	Walk	Cycle	Train
Number of students	8	6	10	4	2

Draw a pie chart to illustrate these data.

Solution $360 \div 30 = 12$ so each student is represented by 12°.
Table for angles is:

Method of travel	Bus	Car	Walk	Cycle	Train
Angle	$8 \times 12 = 96°$	$6 \times 12 = 72°$	$10 \times 12 = 120°$	$4 \times 12 = 48°$	$2 \times 12 = 24°$

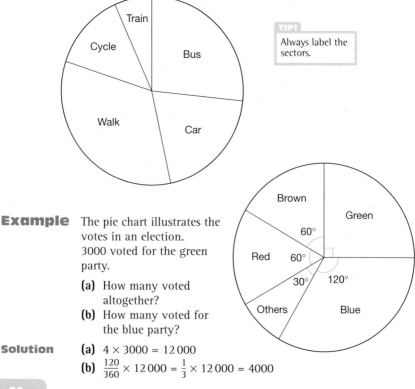

TIP!
Always label the sectors.

Example The pie chart illustrates the votes in an election.
3000 voted for the green party.

(a) How many voted altogether?
(b) How many voted for the blue party?

Solution (a) $4 \times 3000 = 12\,000$
(b) $\frac{120}{360} \times 12\,000 = \frac{1}{3} \times 12\,000 = 4000$

Practice questions

1   In a survey, 40 people were asked what they had drunk that morning. The
    results are shown in this table.

Drink	Coffee	Tea	Fruit juice	Mineral water
Number of people	18	9	7	6

Draw a pie chart to illustrate this information.

2   The table shows which sports the 180 students in Year 10 chose on games
    afternoon.

Sport	Football	Rugby	Gym	Basketball	Badminton
Number of students	70	30	35	25	20

Draw a pie chart to illustrate this information.

3   Gary conducted a survey on the types of vehicle passing his school one
    morning. The results are shown in this table.

Type of vehicle	Cars	Lorries	Buses	Vans	Motorcycles
Frequency	52	7	5	16	10

Draw a pie chart to illustrate this information.

4   The pie chart illustrates the
    numbers of different types of
    homes on an estate.

    There are 150 bungalows.
    (a) How many detached two-storey
    houses are there?
    (b) How many homes are there
    altogether?
    (c) How many flats are there?

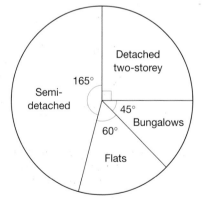

> ❗ **Probabilities are measured on a scale from 0 to 1**
> - 0 means the event is impossible.
> - 1 means the event is certain.
> - When outcomes are equally likely,
>
> $$\text{probability (event A happens)} = \frac{\text{number of ways that A can happen}}{\text{total possible number of outcomes}}$$
>
> **Mutually exclusive events**
> - These cannot happen together.
> - If A, B and C are all the mutually exclusive events that can happen, then $P(A) + P(B) + P(C) = 1$
> - P(A does not happen) = 1 − P(A happens)
>
> > **TIP!**
> > P(A) is a useful shorthand for the probability that A happens.

Example   A bag contains red, blue and green pens and no other colours. A pen is taken out at random.
The probability of taking a red pen is 0·6 and the probability of taking a blue pen is 0·25.
(a) What is the probability of taking a green pen?
(b) If there are 20 pens in the bag, how many of them are red?

Solution   (a) P(green) = 1 − (0·6 + 0·25) = 0·15
(b) 0·6 × 20 = 12

> **TIP!**
> Fractions or decimals (or sometimes percentages) are used for probabilities. Never use ratios or 'in' or 'out of' for probabilities – you will lose marks if you do.

> ❗ **Listing outcomes from two independent events**
> Be systematic to make sure you have shown all the possibilities.

Example   Jenny threw a four-sided and a six-sided dice. The dice were fair.
(a) Draw a diagram to show the possible outcomes.
(b) What was the probability that she threw a total of 8? Give your answer as a fraction in its lowest terms.
(c) What was the probability that she threw a total of 12?

Solution   (a)

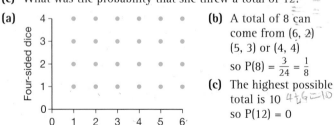

(b) A total of 8 can come from (6, 2) (5, 3) or (4, 4)
so $P(8) = \frac{3}{24} = \frac{1}{8}$

(c) The highest possible total is 10 $4+6=10$
so $P(12) = 0$

Practice questions

1   The probability that the Jones family uses their washing machine today is 0·8. What is the probability that they don't use it today?

2   Three friends are planning an evening out. The probability they will go to the cinema is 0·4. The probability they will hire a video is 0·25. What is the probability they will do something else instead?

3   **(a)** A box contains 10 milk, 4 white and 6 plain chocolates. Jenny takes a chocolate without looking. What is the probability she takes a white one? Give your answer as a fraction in its lowest terms.

(b) The probability of choosing a plain chocolate from another box is $\frac{1}{6}$. This box has 30 chocolates in it. How many of them are plain?

4   A dice is thrown 500 times. Here are the results.

Number on dice	1	2	3	4	5	6
Frequency	52	80	152	90	64	62

(a) Explain why these results suggest the dice is biased.
(b) Find the experimental probability of throwing a 5 with this dice. Give your answer as a decimal.

5   Jim, Kie and Leo go to the theatre. They have seats P1, P2, P3.
(a) List the different ways they can sit.
(b) If they take their tickets at random, what is the probability that Jim and Leo will sit next to each other?

6   Bill threw a fair four-sided dice numbered 1 to 4 and an ordinary six-sided dice. The score for his game was the numbers on the dice multiplied together. Draw a diagram to show the possible scores and find the probability that his score was 12.

7   A restaurant has some special offer two-course meals. These are the choices.
(a) List the possible two-course meals.
(b) From long experience, the chef knows that the probability of a customer choosing lasagne is 0·3.

Main course	Dessert
Shepherd's pie	Ice-cream
Lasagne	Fruit salad
Beef curry	
Macaroni cheese	

What is the probability that a customer doesn't choose lasagne?
(c) Yesterday, the chef served 80 customers with these special offer meals. Estimate how many portions of lasagne he served.

8   Joe has a bag of 50 sweets. 18 of them are red, 15 are yellow, 6 are green, 5 are orange and the rest are purple. Joe takes a sweet without looking. Find the probability that his sweet is
(a) red or orange
(b) black
(c) purple.

TREE DIAGRAMS

> - When the probability of an event happening is p, the probability of it not happening is $1 - p$.
> - Two events can be shown on a tree diagram.

Example Kenny's journey to work involves catching a train and then a bus. On any day in winter, the probability that the train is late is $\frac{1}{6}$.

Independently, the probability that Kenny misses the bus is $\frac{4}{5}$.

Draw and label a tree diagram showing all the possible outcomes for Kenny's journey to work on any winter day.

Solution

```
      Train            Bus
                   Miss 4/5
      Late 1/6
                   Catch 1/5
                   Miss 4/5
      Not late 5/6
                   Catch 1/5
```

TIP!
If P (late) is $\frac{1}{6}$ then
P(not late) is $1 - \frac{1}{6} = \frac{5}{6}$

TIP!
If P(miss) is $\frac{4}{5}$ then
P(catch) is $1 - \frac{4}{5} = \frac{1}{5}$

Example Bill is taking a test. The test has a theory part and a practical part and these are independent.

(a) Complete the tree diagram to show all the possible outcomes.

(b) Calculate the probability that Bill passes only one part of the test.

```
      Theory            Practical
                     Pass 0·7
      Pass 0·8
                     Fail –
                     Pass 0·7
      Fail –
                     Fail –
```

Solution (a)

```
      Theory            Practical
                     Pass 0·7
      Pass 0·8
                     Fail 0·3
                     Pass 0·7
      Fail 0·2
                     Fail 0·3
```

(b) Pass theory, fail practical
$0·8 \times 0·3 = 0·24$

Fail theory, pass practical
$0·2 \times 0·7 = 0·14$

Pass only one part
$0·24 + 0·14 = 0·38$

Practice questions

1   In a hospital, the probability that any new-born baby is male is 0·5 and, independently, the probability that it is left-handed is 0·15.
Draw and label a tree diagram showing all the possible outcomes for the next baby born in the hospital.

2   Rebecca has an old car that she uses to travel to college. The probability that she has to stop at the traffic lights is $\frac{1}{5}$.
The car will break down with probability $\frac{1}{25}$.
These events are independent.
  (a) Complete the tree diagram to show all the possible outcomes.
  (b) What is the probability that, on a single journey to college, Rebecca has to stop at the traffic lights and the car does not break down?

3   It is estimated that, when playing at home, a football team have a 0·6 chance of winning.
  (a) Complete the tree diagram to show the possible outcomes of the next two home games.
  (b) Calculate the probability that they win only one of the games.

4   There are five yellow balls, three blue balls and two green balls in a bag. Melinder chooses a ball, makes a note of its colour and puts it back in the bag. She then chooses another ball.
  (a) Draw a tree diagram to show all the possible outcomes.
  (b) Calculate the probability that
    (i)   both balls are the same colour.
    (ii) she does not choose a yellow ball.

5   Every Sunday morning Keira does a paper round. The probability that she makes a wrong delivery is $\frac{1}{25}$. The probability that she sleeps in and makes the deliveries late is $\frac{2}{15}$.
  (a) Draw a tree diagram to show all the possible outcomes.
  (b) Calculate the probability that all the deliveries are on time and correct.

6   Estelle throws two ordinary fair dice. She is interested in the probability of getting a prime number on one of them and a square number on the other.
  (a) Draw a tree diagram to show all the possible outcomes.
  (b) Calculate the probability that she gets a prime number on one of them and a square number on the other.

> ### Changing the subject of a simple formula
> - Use the techniques for solving equations that were used in Topic 11.
> - Do the same operation to each side of the formula.
> - Work towards getting the letter you want on its own.

Example Rearrange the formula $y = 5x - 3$ to make x the subject.

Solution

$$y = 5x - 3 + 3$$

Add 3 to each side
$$y + 3 = 5x$$

Divide each side by 5
$$\frac{y + 3}{5} = x$$

Rewrite with x on the left
$$x = \frac{y + 3}{5}$$

Example Make h the subject of the formula $V = \pi r^2 h$.

Solution

$$V = \pi r^2 h$$

Divide each side by πr^2
$$\frac{V}{\pi r^2} = h$$

Rewrite with h on the left
$$h = \frac{V}{\pi r^2}$$

> ### Solving simple inequalities
> - Use the techniques for solving equations that were used in Topic 11.
> - Collect the letter terms on the side which has more of them.

Example Solve $5x - 2 > 3x + 1$. Represent the solution on a number line.

Solution

$$5x - 2 > 3x + 1$$

Add 2 to each side
$$5x > 3x + 3$$

Subtract $3x$ from each side
$$2x > 3$$

Divide each side by 2
$$x > 1{\cdot}5$$

```
 ┌──┬──┬──┬──┬──┬──┬─⊙──┬──┬──┬──→
-5  -4  -3  -2  -1   0   1   2   3   4   5
```

> **TIP!**
>
> Use an open circle to represent > or < and use a closed circle for $\geqslant$ or $\leqslant$ (when you need to include the number itself).

Example Solve $3x < 10 - x$.

Solution

$$3x < 10 - x$$

Add x to each side
$$4x < 10$$

Divide each side by 4
$$x < 2{\cdot}5$$

> ### Expanding two pairs of brackets
> - Each term in the first bracket needs to be multiplied by each term in the second bracket.
> - Simplify the answer by collecting like terms.
> - You can use a grid method or FOIL (First × First, Outer × Outer, Inner × Inner, Last × Last)

Practice questions

1 Rearrange these formulae so that the letter shown becomes the subject.

(a) $C = \pi d$ new subject d

(b) $A = \frac{1}{2}bh$ new subject b

(c) $y = 5 + 2x$ new subject x

(d) $A = 40 + 4t$ new subject t

(e) $C = 4a - 5$ new subject a

(f) $d = \frac{m}{v}$ new subject m

(g) $P = 2W + 2L$ new subject W

(h) $v = u + at$ new subject u

(i) $v = u + at$ new subject a

(j) $V = abh$ new subject h

2 Solve these inequalities.

(a) $5x \geqslant 15$ (b) $2x - 1 < 4$

(c) $6 + x > 2$ (d) $9 + 2x \leqslant 10$

(e) $\frac{1}{2}x > 4$ (f) $3x - 5 > 10$

(g) $5x > 3x + 6$ (h) $4 + 2x < 10x$

(i) $5a - 3 \geqslant 7$ (j) $5 < 3c + 8$

3 Expand and simplify these.

(a) $(x + 2)(x + 3)$ (b) $(x + 5)(x + 1)$

(c) $(x - 2)(x + 3)$ (d) $(x + 2)(x - 3)$

(e) $(x - 7)(x + 7)$ (f) $(y + 2)(y - 2)$

(g) $(y - 3)(y - 4)$ (h) $(t - 6)(t + 1)$

(i) $(a - 3)(a - 5)$ (j) $(x + 2)(x - 4)$

Example Expand and simplify $(y + 3)(y - 5)$.

Solution $(y + 3)(y - 5) = y^2 + 3y - 5y - 15$
$= y^2 - 2y - 15$

x	y	$+3$
y	y^2	$+3y$
-5	$-5y$	-15

Example Expand and simplify $(a - 4)(a - 2)$.

Solution   F O I L
$(a - 4)(a - 2) = a^2 - 2a - 4a + 8$
$= a^2 - 6a + 8$

TIP!
You may want to draw the arrows as shown and remember a smiley face shape.

Repeated percentage increase
- The quick method for increasing by a percentage is to use a multiplier.
- Examples of multipliers are given in this table.

Increase	5%	12%	$17\frac{1}{2}\%$	150%
Multiplier	1·05	1·12	1·175	2·5

- To increase by 5% four times, say, multiply by 1·05 × 1·05 × 1·05 × 1·05, or by 1·05^4.

Example The population of a country increased by 15% every 10 years. The population in 1960 was 15 600 000. Find the population in 2000. Give the answer to the nearest thousand.

Solution Multiplier is 1·15, time is 40 years = 4 × 10 years.

Population in 2000 is 15 600 000 × 1·15 × 1·15 × 1·15 × 1·15

$$= 15\,600\,000 \times 1·15^4$$

$$= 27\,284\,000 \text{ to the nearest thousand.}$$

TIP!

To multiply by 1·15^4 use the power button (y^r or ^ or x^y) on your calculator.

To do this calculation press 1 5 6 0 0 0 0 0 × 1 · 1 5 y^r 4 =

Compound interest
- Compound interest is an example of a repeated percentage increase.
- Each year the amount at the start of the year increases by a given percentage so a multiplier is used.

Repeated percentage decrease
- The quick method for decreasing by a percentage is to use a multiplier.
- Examples of multipliers are given in this table.

Decrease	5%	15%	23%	60%
Multiplier	0·95	0·85	0·77	0·40

- To decrease by 5% four times, say, multiply by 0·95 × 0·95 × 0·95 × 0·95, or by 0·95^4.
- When the value of an item goes down repeatedly, this is called **depreciation**.

Practice questions

1   The population of a colony of bacteria increases by 2% per hour. If there were 3000 bacteria at 12 noon, how many will there be at 6 pm?

In the following seven questions give the answers to the nearest pound.

2   The value of an antique increases by 60% every 5 years. If it was worth £400 in 1990, how much will it be worth in 2020?

3   House prices increased by 14% per year from 2002 to 2007. If a house cost £71 000 in 2002, what did it cost in 2007?

4   Sarah invested £10 000 on 1st January 2004 at 5% compound interest. How much was the investment worth on 1st January 2007?

5   Rajvee invested £20 000 on 1st May 1997 at $4\frac{1}{2}$% compound interest. What was this investment worth on 1st May 2007?

6   When he was born, Jack's grandmother put £500 in a bank account for him at 4% compound interest. How much interest had this earned by his eighteenth birthday?

7   Lisa bought a car in 2003 for £8000. If depreciation was 15% per year, what was the car worth in 2008?

8   The population of a certain species of bird fell by 25% every 10 years from 1950 to 2000. If the population was 3 million in 1950, what was the population in 2000?

Example   Colin invested £5000 in the bank on 1st January 2005. If the compound interest rate is 6%, what will Colin's investment be worth on 1st January 2012? Give the answer to the nearest penny.

Solution   The investment is for 7 years and the multiplier is 1·06.
Amount = 5000 × 1·06 × 1·06 × 1·06 × 1·06 × 1·06 × 1·06
         × 1·06
       = 5000 × 1·06^7 = £7518·15

Example   Sanjay bought a new car in 2006 for £12 500. Its value reduces by 12% each year. What will it be worth in 2010? Give the answer to the nearest pound.

Solution   Multiplier is 0·88, number of years = 4.
Value in 2010 is   12 500 × 0·88 × 0·88 × 0·88 × 0·88
                 = 12 500 × 0·88^4 = £7496.

> In **standard index form,** numbers are always written as $A \times 10^n$ where A is a number between 1 and 10 and n is the number of places the decimal point has been 'moved'.

Example The distance from the Earth to the Sun is approximately 150 000 000 km. Write 150 000 000 in standard index form.

Solution A is 1.5
and n is 8
1.5×10^8

> **TIP!** The number part has to be between 1 and 10.

Example Write the following numbers as ordinary numbers.

> **TIP!** Imagine the decimal point between the 1 and the 5 and count the spaces to the right.

(a) 4.7×10^6 (b) 6.3×10^{-5}

Solution (a) $4.7 \times 1\,000\,000 = 4\,700\,000$ (b) $6.3 \times \dfrac{1}{100\,000} = 0.000\,063$

> When manipulating numbers given in standard index form, deal with the 'number' part first and the 'index' part second.

Example Work out $(7 \times 10^4) \times (9 \times 10^5)$, expressing the answer in standard form.

Solution $7 \times 9 = 63$
$10^4 \times 10^5$ is 10^9

> **TIP!** Deal with the 'number' parts first.

> **TIP!** Remember to add the indices when multiplying.

63×10^9 appears to be the answer but the number part **must** be between 1 and 10.
$63 \times 10^9 = (6.3 \times 10) \times 10^9$
$= 6.3 \times 10^{10}$

> **TIP!** Remember that 10 is really 10^1.

> When adding or subtracting numbers in standard index form, remember to change into 'ordinary' numbers **before** doing the calculation and then change the answer into standard index form at the end.

Example Work out these, expressing the answers in standard form.

(a) $(4 \times 10^6) + (5 \times 10^4)$ (b) $(6 \times 10^5) - (7.8 \times 10^4)$

Solution (a) $(4 \times 10^6) = 4\,000\,000$ (b) $(6 \times 10^5) = 600\,000$
$(5 \times 10^4) = 50\,000$ $(7.8 \times 10^4) = 78\,000$
$4\,000\,000 + 50\,000 = 4\,050\,000$ $600\,000 - 78\,000 = 512\,000$
$4\,050\,000 = 4.05 \times 10^6$ $512\,000 = 5.12 \times 10^5$

Practice questions

1   Write these numbers in standard index form.

    (a) 750 000   (b) 8 200 000   (c) 76 000   (d) 56 400

    (e) 0·000 006   (f) 0·029   (g) 0·000 78   (h) 0·659

2   Write these numbers as ordinary numbers.

    (a) 7×10^3 (b) 9×10^5 (c) $6·8 \times 10^4$ (d) $7·4 \times 10^6$

    (e) $8 \div 10^4$ (f) $6 \div 10^3$ (g) $7·4 \div 10^5$ (h) $4·9 \div 10^4$

3   Work these out, expressing the answers in standard index form.

    (a) $(4 \times 10^3) \times (7 \times 10^4)$ (b) $(6 \times 10^7) \times (9 \times 10^4)$

    (c) $(6 \times 10^8) \div (8 \times 10^3)$ (d) $(5 \times 10^8) \div (7 \times 10^3)$

    (e) $(3·4 \times 10^4) \times (6 \times 10^5)$ (f) $(7 \times 10^3) \times (3·6 \times 10^4)$

    (g) $(8 \times 10^6) \div (5·2 \times 10^3)$ (h) $(7·3 \times 10^5) \div (6·7 \times 10^3)$

4   Work these out, expressing the answers in standard index form.

    (a) $(6 \times 10^4) + (5 \times 10^3)$ (b) $(4 \times 10^6) + (7 \times 10^5)$

    (c) $(4 \times 10^4) + (5·7 \times 10^5)$ (d) $(7·6 \times 10^4) + (5 \times 10^3)$

    (e) $(9 \times 10^5) - (6 \times 10^4)$ (f) $(7 \times 10^5) - (5 \times 10^4)$

    (g) $(6·2 \times 10^6) - (4 \times 10^5)$ (h) $(4 \times 10^6) - (7·3 \times 10^5)$

5   The diameter of the Sun is approximately 1 390 000 km. Write this number using standard index form.

6   The diameter of a helium atom is about 0·000 000 000 1 m. Write this number using standard index form.

7   The rotational speed of the Earth is approximately $1·07 \times 10^5$ km/h. Write this speed as an ordinary number.

8   The fastest ever manned vehicle was the Apollo 10 capsule on its return to Earth in 1969. The speed recorded on re-entry to the Earth's atmosphere was approximately $2·48 \times 10^4$ m.p.h. Write this speed as an ordinary number.

Topic 1 Percentages

1 (a) £6 (b) £6 (c) £22·50 (d) £144 2 (a) 25% (b) 15% (c) 30% (d) 60% (e) 5% 3 225 4 87·5%
5 £73·60 6 £28·20 7 £960 8 68% 9 £7800 10 70·8% 11 25% 12 20% 13 £81 14 (a) Profit 3%
(b) Loss 12·5% 15 £28 529 16 £11481·75 17 £16 390·91

Topic 2 Fractions

1 (a) $\frac{3}{4}$ (b) $\frac{1}{4}$ (c) $\frac{3}{5}$ (d) $\frac{3}{10}$ 2 (a) $\frac{7}{8}$ (b) $\frac{2}{3}$ (c) $\frac{11}{12}$ (d) $\frac{9}{10}$ (e) $\frac{7}{10}$ (f) $\frac{19}{20}$ (g) $\frac{9}{10}$ (h) $\frac{17}{15} = 1\frac{2}{15}$

3 (a) $\frac{1}{2}$ (b) $\frac{1}{3}$ (c) $\frac{1}{2}$ (d) $\frac{1}{4}$ (e) $\frac{1}{28}$ (f) $\frac{3}{20}$ (g) $\frac{1}{3}$ (h) $\frac{1}{18}$ 4 (a) $3\frac{5}{6}$ (b) $2\frac{1}{6}$ (c) $9\frac{9}{10}$ (d) $4\frac{5}{12}$ (e) $1\frac{1}{5}$

(f) $2\frac{17}{30}$ (g) $6\frac{4}{21}$ (h) $\frac{19}{20}$ 5 (a) $3\frac{3}{5}$ (b) $7\frac{1}{3}$ (c) $\frac{2}{25}$ (d) $\frac{21}{80}$ (e) $\frac{1}{8}$ (f) $\frac{4}{15}$ (g) $\frac{1}{8}$ (h) $\frac{1}{6}$ 6 (a) $\frac{1}{6}$ (b) 12

(c) $\frac{5}{8}$ (d) $\frac{18}{35}$ (e) $\frac{3}{10}$ (f) $1\frac{2}{3}$ (g) $\frac{2}{3}$ (h) $1\frac{1}{5}$ 7 (a) $\frac{5}{6}$ (b) $\frac{2}{5}$ (c) $\frac{1}{18}$ (d) 1

Topic 3 Non-calculator work

1 (a) 145·2 (b) 1730 (c) 2·46 (d) 0·27 (e) 460 (f) 0·0172 (g) 293 200 (h) 0·836 (i) 2 869 000 2 7300 g
3 6·34(0) kg 4 2780 ml 5 0·738 litres 6 3·7 mm 7 780 cm 8 4·35 m 9 (a) 44·62 (b) 70·823 (c) 19·59
(d) 27·72 10 (a) 45·3 (b) 0·563 (c) 0·0196 (d) 5·23 11 18·3 cm 12 (a) 0·15 (b) 0·08 (c) 0·05 (d) 0·04
13 (a) 1800 (b) 14 000 (c) 300 000 14 416 weeks 15 6939 days 16 (a) £6·30 (b) £2·80 (c) £1·05
(d) £4·20 (e) 0·75 m or 75 cm (f) £1·98

Topic 4 Ratio

1 £14 : £21 : £21 2 15 cm 3 1·5 m 4 28 cm 5 2·5 cm 6 4000 7 16 girls 8 19·5 cm 9 £6 : £9 : £15
10 75 cm 11 (a) 3 : 8 (b) 160 g 12 7·2 cm 13 £360 14 24 oranges 15 36 cm

Topic 5 Direct proportion, best buy and money problems

1 $x = 8$, $y = 25$ 2 $A = 4$, $W = 20$ 3 (a) $d = 43g$ (b) 344 miles (c) 11·6 gallons 4 (a) £3·20 (b) 125 miles
5 (a) $y = 4x^2$ (b) $y = 100$ (c) $x = 10$ 6 (a) $d = 5t^2$ (b) 80 m (c) 5 seconds 7 400 g size 8 £6·60
9 £35·70 10 £1595 11 Scheme A: £1692, £282; Scheme B: £2538, £164·50

Topic 6 Primes, HCF, LCM and power laws

1 (a) 2^4 (b) $2^2 \times 3^2$ (c) 5×3^2 (d) $2 \times 3 \times 5 \times 7$ (e) $2^2 \times 3 \times 5^2$ 2 (a) 7 (b) 8 (c) 13 (d) 1 (e) 2 (f) 9
(g) 17 3 (a) 252 (b) 140 (c) 216 (d) 1344 (e) 280 (f) 1386 (g) 168 4 288 5 1260 6 (a) 3^6 (b) a^9
7 (a) 5^2 (b) 9^2 (c) k 8 (a) p^6 (b) 7^{30} 9 (a) x^9 (b) y^9 (c) 2^{15} 10 (a) p^{22} (b) $120t^{11}$ 11 (a) $5x^3$ (b) y^5
12 (a) $32m^7$ (b) 1 (c) $84t^4$ 13 (a) 7^6 (b) $8x^3$ (c) $8p^{12}$

Topic 7 Estimation and rounding

1 (a) 1250 (b) 1300 (c) 7000 (d) 3600 (e) 7·2 (f) 0·03 (g) 1·67 (h) 0·063 2 11 000 3 (a) $50 \times 80 = 4000$
(b) $200 \times 60 = 12\,000$ or 10 000 (c) $\frac{20 \times 30}{50} = 12$ or 10 (d) $\frac{200 \times 20}{40} = 100$ (e) $\frac{60 \times 500}{30 \times 50} = 20$

4 (a) $\frac{2 \times 20}{5} = 8$ (b) $\frac{3 + 6}{9 - 7} = \frac{9}{2} = 4 \cdot 5$ or 4 or 5 (c) $\frac{20 \times 20}{6 - 2} = \frac{400}{4} = 100$ (d) $\frac{70 \times 50}{80 - 50} = \frac{3500}{30} = 116 \cdot 6 = 120$ or 100

5 (a) 7·7 (b) 3·8 (c) 122·0 (d) 110·7 6 (a) $400 \times 0\cdot4 = 160$ or 200 (b) $0\cdot8 \times 0\cdot06 = 0\cdot048 = 0\cdot05$
(c) $0\cdot5 \div 8 = 0\cdot0625 = 0\cdot06$ (d) $(2 \times 600) \div (70 + 20) = 1200 \div 90 = 13\cdot3 = 13$ or 10 7 (a) $200 \times 200 \div 0\cdot8 = 40\,000$
$\div 0\cdot8 = 50\,000$ (b) $(500 + 700) \div 0\cdot8 = 1200 \div 0\cdot8 = 1500$ (c) $(40 \times 40) \div (0\cdot8 \times 6) = 1600 \div 4\cdot8 = 1600 \div 5 = 320$
$= 300$ (d) $(3 \times 600) \div (7 \times 0\cdot7) = 1800 \div 4\cdot9 = 1800 \div 5 = 360$ or 400 8 $\frac{200}{4} = 50$ m.p.h. 9 (a) Estimate of
population density = $3217 \div 52\cdot9 = 3000 \div 50 = 60$ (b) population density = 60·8 10 354 cm³

Topic 8 Using a calculator

1 5·78 2 13 3 5·32 4 2·88 5 28·67 6 −37·7 7 0·27 8 $9\frac{5}{28}$ 9 33·08 10 $1\frac{1}{5}$ 11 39·06 12 15·85
13 1·57 14 $13\frac{3}{20}$ 15 309·27 16 5·00 17 $-8\cdot15 \times 10^{-3}$ 18 7·79 19 2130·01 20 1·35

Topic 9 Algebraic manipulation

1 (a) $7x + 12$　(b) $5x - 9y$　(c) $3a + 2b$　(d) $7x - 8y + 5$　(e) $4x^2 - 4x + 4$　(f) $-3x^2 - 9x - 5$　2 (a) $12a$　(b) $10b^2$
(c) $18x^2$　(d) $10y^3$　(e) $18ab$　(f) $30abc$　3 (a) 2　(b) $5a$　(c) $3x$　(d) $5x^2$　(e) $4y$　4 (a) $5a + 20$　(b) $7x - 14$
(c) $x^2 + 3x$　(d) $ab - 3a$　(e) $3a + 3b + 3c$　5 (a) $2(x + 2)$　(b) $3(a - 4)$　(c) $x(x + 5)$　(d) $y(x - 3)$
(e) $5(x + y - z)$　6 (a) $8x + 11$　(b) $7a - 5$　(c) $2x - 18$　7 (a) x^3　(b) $6a^3$　8 (a) $2x(x + 3)$　(b) $5a(a - 2)$
(c) $3x(2x + 1)$　(d) $2a(2a - 3b)$　(e) $5y(x + 2z)$

Topic 10 Substitution

1 (a) 10　(b) –12　(c) 3　(d) 11　(e) –24　(f) –2　(g) 54　(h) 45　(i) 32　(j) 10　2 (a) –12　(b) –6　(c) 8
(d) 10　(e) 51　3 (a) 13　(b) 1　(c) –7　4 (a) 13　(b) –2　(c) –11　5 (a) 8　(b) –1　(c) 15

6

y	7	4	3	4	7	12	19

7 10·5　8 (a) 15　(b) 5　9 20　10 1, 8, 15　11

y	8	6	4	2	0	–2

12 $4\frac{1}{2}$ cm^2　13 32　14 6·5　15 18 cm　16 24·5 cm to 1 d.p.　17 84·9 cm^2 to 1 d.p.　18 4·5　19 43·74 cm^2

Topic 11 Linear equations

1 5　2 6　3 2　4 $\frac{4}{5}$　5 $2\frac{1}{3}$　6 $7\frac{1}{2}$　7 6　8 10　9 5　10 2　11 5　12 7　13 -17·5　14 -56　15 1　16 3　17 6
18 2　19 1　20 $\frac{3}{2}$　21 2　22 7　23 8　24 2　25 –3　26 –3　27 –7

Topic 12 Trial and improvement

1 3·8, 4·8, 6·8　2 3·3 cm, 8·3 cm, 18·3 cm　3 3·25 cm, 6·25 cm, 6·25 cm　4 5·5 cm, 11 cm, 16·5 cm　5 2·6　6 1·9
7 -2·7　8 3·41　9 3·43　10 2·82

Topic 13 Drawing graphs

1 (a) $y = 3x + 2$

x	–3	–2	–1	0	1	2
y	–7	–4	–1	2	5	8

2 (a) $y = 2 - x$

x	–2	–1	0	1	2	3
y	4	3	2	1	0	–1

(b)

(b)

3 (a) $2y = x + 4$

x	–4	–3	–2	–1	0	1	2
y	0	0·5	1	1·5	2	2·5	3

4 (a) $y = 2x - 1$

x	–2	0	4
y	–5	–1	7

(b)

(b) $x = 2·5$ when $y = 4$

5 (a) $y = 2x + 3$

x	–2	0	4
y	–1	3	11

(b) $2y = 8 - x$

x	–2	0	4
y	5	4	2

(c) $x = 0·4$, $y = 3·8$ (accept ± 0·2)

6 (a)

x	–3	–2	–1	0	1	2	3
y	3	–2	–5	–6	–5	–2	3

(b)

(c) $x = -2·4$ and $+2·4$ (accept ± 0·1)

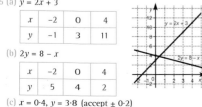

7 (a)

x	-3	-2	-1	0	1	2	3	4
y	6	0	-4	-6	-6	-4	0	6

8 (a)

x	-3	-2	-1	0	1	2
y	7	-1	-5	-5	-1	7

(b), (c) (i)

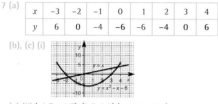

(b)

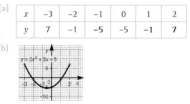

(c) (ii) (-1·7, -1·7), (3·7, 3·7) (accept ± 0·2)

(c) x = -2·2 and 1·2 (accept ± 0·1)

Topic 14 Angle properties

1 a = 45°, angles on a straight line   2 b = 60°, angles on a straight line   3 c = 75°, angles on a straight line
4 d = 120°, angles around a point   5 e = 110°, angles around a point   6 g = 75°, angles in a triangle
7 h = 25°, angles in a triangle   8 i = 99°, angles in a triangle   9 j = 16°, vertically opposite angles k = 164°,
angles on a straight line   10 l = 41°, base angles of an isosceles triangle m = 98°, angles in a triangle
11 n = 32°, angles in a triangle, base angles of an isosceles triangle   12 p = 110°, angles on a straight line q = 40°,
angles in a triangle   13 r = 83°, angles in a triangle s = 97°, angles on a straight line   14 t = 45°, angles in a
quadrilateral u = 45°, vertically opposite angles   15 a = 65°, corresponding angles b = 69°, allied angles
16 c = 30°, alternate angles d = 130°, allied angles   17 e = 72°, alternate angles f = 45°, corresponding angles
18 g = 50°, alternate angles h = 55°, allied angles

Topic 15 Circle angle facts

1 (a) x = 50°, y = 80°, z = 100°   (b) v = 60°, x = 30°, y = 120°, z = 60°   2 (a) x = 54°, y = 108°   (b) a = 50°,
b = 100°   3 (a) x = 135°   (b) x = 260°, y = 100°, z = 130°   4 (a) a = 45°, b = 45°   (b) a = 60°, b = 30°
5 m = 90°, n = 50°   6 p = 63°   7 r = 164°

Topic 16 Constructions and loci

1 Check your angles. Bisected angles: (a) 30°   (b) 17° or 18°   (c) 65°.   2 Check your diagrams. The
perpendicular bisectors should be at 90° to the line in each case and divide the lines into   (a) 5 cm   (b) 2·5 cm
(c) 5·7 cm or 5·8 cm.   3 Check your diagram. The perpendicular bisector should be at 90° to the line.   4 Check your diagram. The perpendicular bisector should be at 90° to the line.   5 Circle, centre A, radius 3 cm
6 Bisector of angle CAB   7 Perpendicular bisector of AB

8

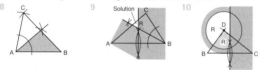

9   Solution

10

Topic 17 Transformations

1

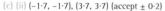

(a) A:(1, -1), (2, -1), (1, -3)   (b) B:(1, 5), (2, 5), (1, 3)   (c) C:(-1, 1), (-1, 2), (-3, 1)
(d) D:(-1, -3), (-2, -3), (-1, -5)   2 (a) E:(2, 2), (4, 2), (2, 6)   (b) F:(-3, -1), (0, -1), (-3, 5)
(c) G:(3, -4), (4, -4), (3, -2)   3 (a) Reflection in the y-axis   (b) Translation of 3 units to
the right and one up   (c) Translation of 3 units to the left and one down   (d) Reflection in
y = -1   (e) Rotation through 90° clockwise about (0. 0)   (f) Reflection in y = x
4 (a) Enlargement with scale factor 3 and centre (0, 1)   (b) Enlargement with scale factor
$\frac{1}{3}$ and centre (0, 1)   (c) Rotation through 180° about (1·5, -1·5)   5 (a) J:(2, 1), (3, 1), (2, 3)
(b) K:(3, -4), (3, -5), (5, -4)   (c) L:(-2, -4), (-2, -6), (-6, -4)   (d) M:(1, -6), (3, -6), (1, -2)

Topic 18 Pythagoras

1 $x = 13$ cm   2 $x = 8.1$ cm   3 $x = 6$ cm   4 $x = 24$ cm   5 $x = 9.2$ m   6 $x = 16.5$ m   7 15 cm   8 11.5 cm   9 5.8 cm
10 3.8 m   11 6 cm   12 30.5 m   13 7.2 units   14 6.2 m   15 61.0 m

Topic 19 Area, volume and perimeter

1 $W = 6$ cm, $A = 60$ cm^2 2 $P = 24$ cm, $A = 24$ cm^2 3 54 cm^2 4 0.6 m^2 5 900 cm^2 6 25 cm^2 7 $V = 60$ cm^3,
$S = 104$ cm^2 8 $V = 36$ cm^3, $S = 84$ cm^2 9 4.7 m^3 10 6 cm

Topic 20 Speed and other compound units

1 (a) 157 miles   (b) The coach stopped for half an hour.   (c) 58 m.p.h.   2 (a) 32 km/h   (b) Line joining (0, 120)
to (2, 0). They pass at 56 km from Jenny's home.   3 45 m.p.h.   4 5208 km   5 $2\frac{1}{2}$ hours   6 142 km/h
7 Sarah 4 km/h, Mike 3.84 km/h, Sarah fastest   8 410 m   9 10 hours 35 minutes   10 63 people/km^2
11 40.9 m.p.g.   12 8.5 g/cm^3

Topic 21 Scatter diagrams

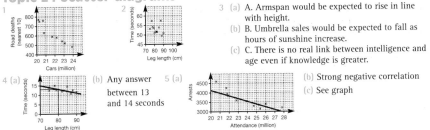

1   2   3 (a) A. Armspan would be expected to rise in line
with height.
(b) B. Umbrella sales would be expected to fall as
hours of sunshine increase.
(c) C. There is no real link between intelligence and
age even if knowledge is greater.

4 (a)   (b) Any answer   5 (a)   (b) Strong negative correlation
between 13   (c) See graph
and 14 seconds

Topic 22 Cumulative frequency and box plots

1 (a) 200 girls   (b) 100 girls at 160, so 200 – 100 = 100 girls between 160 cm and 170 cm   (c) 162 cm
2 Both towns have virtually the same range; the interquartile range for A is greater than for B; the median is
greater for B; 50% of the population of town A are 36 years old or less; 50% of the population for town B are 58
years old or less.   3 Range of heights less for boys; interquartile range is smaller for girls; 50% of girls are
between 150 cm and 175 cm tall; median height for girls is higher.
4 (a) 43   43   45   45   46   46   47   48   48   49   50   51   53   54   56   58   61   62   75   79
(b) Median = 49.5 min, Lower quartile = 46 min, Upper quartile = 57 min

Topic 23 Frequency diagrams and polygons

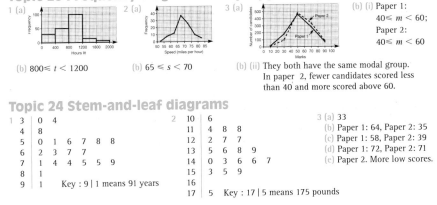

1 (a)   2 (a)   3 (a)   (b) (i) Paper 1:
$40 \leqslant m < 60$;
Paper 2:
$40 \leqslant m < 60$

(b) $800 \leqslant t < 1200$ (b) $65 \leqslant s < 70$ (b) (ii) They both have the same modal group.
In paper 2, fewer candidates scored less
than 40 and more scored above 60.

Topic 24 Stem-and-leaf diagrams

1 3 | 0   4
  4 | 8
  5 | 0   1   6   7   8   8
  6 | 2   3   7   7
  7 | 1   4   4   5   5   9
  8 | 1
  9 | 1   Key : 9 | 1 means 91 years

2 10 | 6
  11 | 4   8   8
  12 | 2   7   7
  13 | 5   6   8   9
  14 | 0   3   6   6   7
  15 | 3   5   9
  16 |
  17 | 5   Key : 17 | 5 means 175 pounds

3 (a) 33
(b) Paper 1: 64, Paper 2: 35
(c) Paper 1: 58, Paper 2: 39
(d) Paper 1: 72, Paper 2: 71
(e) Paper 2. More low scores.

4 (a)

14	1	2	5						
15	1	1	2	3	4	6	6	9	9
16	0	0	1	2	3	4	7	9	
17	1	1	2	4	7	8	9		
18	1	6	8	Key : 18	1 means 181 cm				

(b) 150 to 159   (c) $33\frac{1}{3}\%$ (d) $\frac{3}{30}$ or equivalent

Topic 25 Pie charts

1 Pie chart with angles: Coffee 162°; Tea 81°; Fruit juice 63°; Mineral water 54°. 2 Pie chart with angles: Football 140°; Rugby 60°; Gym 70°; Basketball 50°; Badminton 40°. 3 Pie chart with angles: Cars 208°; Lorries 28°; Buses 20°; Vans 64°; Motorcycles 40°. 4 (a) 300 (b) 1200 (c) 200

Topic 26 Basic probability

1 0·2 2 0·35 3 (a) $\frac{1}{5}$ (b) 5 4 (a) There are far more 3s than any other score. (b) 0·128 5 (a) JKL, JLK, KJL, KLJ, LJK, LKJ (b) $\frac{4}{6}$ or $\frac{2}{3}$ 6 $\frac{1}{8}$ 7 (a) SI, SF, LI, LF, BI, BF, MI, MF (b) 0·7 (c) 24 8 (a) 0·46 (b) 0 (c) 0·12

Topic 27 Tree diagrams

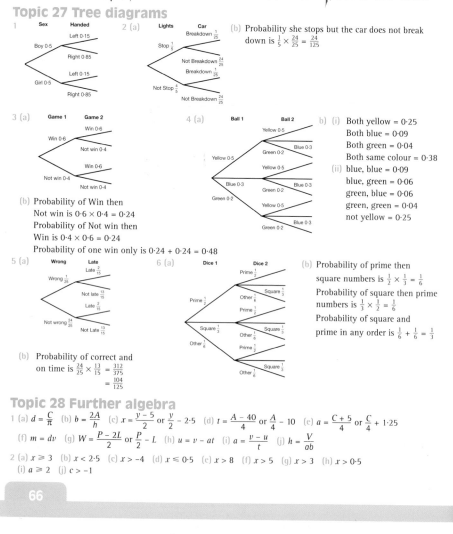

1

2 (a) (b) Probability she stops but the car does not break down is $\frac{1}{5} \times \frac{24}{25} = \frac{24}{125}$

3 (a)

4 (a) b) (i) Both yellow = 0·25
Both blue = 0·09
Both green = 0·04
Both same colour = 0·38
(ii) blue, blue = 0·09
blue, green = 0·06
green, blue = 0·06
green, green = 0·04
not yellow = 0·25

(b) Probability of Win then Not win is $0·6 \times 0·4 = 0·24$
Probability of Not win then Win is $0·4 \times 0·6 = 0·24$
Probability of one win only is $0·24 + 0·24 = 0·48$

5 (a)

(b) Probability of correct and on time is $\frac{24}{25} \times \frac{13}{15} = \frac{312}{375} = \frac{104}{125}$

6 (a) (b) Probability of prime then square numbers is $\frac{1}{2} \times \frac{1}{3} = \frac{1}{6}$
Probability of square then prime numbers is $\frac{1}{3} \times \frac{1}{2} = \frac{1}{6}$
Probability of square and prime in any order is $\frac{1}{6} + \frac{1}{6} = \frac{1}{3}$

Topic 28 Further algebra

1 (a) $d = \frac{C}{\pi}$ (b) $b = \frac{2A}{h}$ (c) $x = \frac{y-5}{2}$ or $\frac{y}{2} - 2·5$ (d) $t = \frac{A-40}{4}$ or $\frac{A}{4} - 10$ (e) $a = \frac{C+5}{4}$ or $\frac{C}{4} + 1·25$
(f) $m = dv$ (g) $W = \frac{P-2L}{2}$ or $\frac{P}{2} - L$ (h) $u = v - at$ (i) $a = \frac{v-u}{t}$ (j) $h = \frac{V}{ab}$

2 (a) $x \geqslant 3$ (b) $x < 2·5$ (c) $x > -4$ (d) $x \leqslant 0·5$ (e) $x > 8$ (f) $x > 5$ (g) $x > 3$ (h) $x > 0·5$
(i) $a \geqslant 2$ (j) $c > -1$

3 (a) $x^2 + 5x + 6$ (b) $x^2 + 6x + 5$ (c) $x^2 + x - 6$ (d) $x^2 - x - 6$ (e) $x^2 - 49$ (f) $y^2 - 4$ (g) $y^2 - 7y + 12$
(h) $t^2 - 5t - 6$ (i) $a^2 - 8a + 15$ (j) $x^2 - 2x - 8$

Topic 29 Repeated change and compound interest
1 3378 2 £6711 3 £136 704 4 £11 576 5 £31 059 6 £513 7 £3550 8 711 914

Topic 30 Standard index form
1 (a) 7.5×10^5 (b) 8.2×10^6 (c) 7.6×10^4 (d) 5.64×10^4 (e) 6×10^{-6} (f) 2.9×10^{-2} (g) 7.8×10^{-4}
(h) 6.59×10^{-1} 2 (a) 7000 (b) 900 000 (c) 68 000 (d) 7 400 000 (e) 0.0008 (f) 0.006 (g) 0.000 074
(h) 0.000 49 3 (a) 2.8×10^8 (b) 5.4×10^{12} (c) 7.5×10^4 (d) 7.14×10^4 (e) 2.04×10^{10} (f) 2.52×10^8
(g) 1.54×10^3 (h) 1.09×10^2 4 (a) 6.5×10^4 (b) 4.7×10^6 (c) 6.1×10^5 (d) 8.1×10^4 (e) 8.4×10^5
(f) 6.5×10^5 (g) 5.8×10^6 (h) 3.27×10^6 5 1.39×10^6 6 1×10^{-10} 7 107 000 8 24 800